Astronomer's Computer Companion

ASTRONOMER'S
COMPUTER COMPANION

Jeff Foust *&* Ron LaFon

no starch press
San Francisco

Trademarks

Trademarked names are used throughout this book. Rather than use a trademark symbol with every occurrence of a trademarked name, we are using the names only in an editorial fashion and to the benefit of the trademark owner, with no intention of infringement of the trademark.

PUBLISHER: William Pollock
PROJECT EDITOR: Karol Jurado
COVER DESIGN AND COMPOSITION: Derek Yee Design
COVER PHOTO: Derek Yee
INTERIOR DESIGN: Margery Cantor
COPYEDITORS: Gail Nelson, Judy Ziajka
PROOFREADER: John Carroll
INDEXER: Nancy Humphreys

Telescope on front cover provided courtesy of Celestron International

Distributed to the book trade in the United States and Canada by Publishers Group West, 1700 Fourth Street, Berkeley, California 94710, phone: 800-788-3123 fax: 510-658-1834.

For information on translations or book distributors outside the United States, please contact No Starch Press directly:

No Starch Press
555 DeHaro Street, Suite 250, San Francisco, CA 94107
phone: 415-863-9900; fax: 415-863-9950; info@nostarch.com; www.nostarch.com

Library of Congress Cataloging-in-Publication Data

Foust, Jeff.
        The astronomer's computer companion / Jeff Foust and Ron LaFon.
                p.   cm.
        Includes index.
        ISBN 1-886411-22-0 (acid-free paper) .
        1.   Astronomy–Data processing. 2. Microcomputers.
    I.   LaFon, Ron. II. Title.
QB51.3.E43F68   1997
522'.85–dc21   97-36006

To my mother—J.F.

Table of Contents

Chapter 16
Remote Astronomy over the Internet

Chapter 17
Computers and Amateur Astronomy

Appendix A
Software on the CD-ROM

Acknowledgments

When I first started corresponding with Bill Pollock of No Starch Press in late 1996 about a book project that would become the *Astronomer's Computer Companion,* I don't think I truly understood what I was getting myself into. I know I underestimated the amount of time and effort that goes into writing a book, especially when trying to juggle a multitude of other projects. To make up for that, though, I also underestimated the thrill one gets when one completes a book, or even a single chapter. It's been a fun adventure.

A number of individuals and companies have graciously provided shareware or demo versions of their software that are included on the CD-ROM. I thank Eric Bergman Terrell, David Chandler, Exploration Software, David Irizarry, Larry Kalinowksi, Microprojects, David Nagy, Parallax Multimedia, Procyon Systems, Rainman Software, Stephen Schimpf, Sienna Software, SkyMap Software, and Southern Stars Software, for their fine cooperation.

Bill Pollock and Karol Jurado at No Starch Press provided expert advice, guidance, and encouragement throughout the writing and editing of this book. It is through their tireless efforts that my words and ideas became the polished final product you're reading.

Most importantly, I want to thank my mother, who always encouraged my interest in both astronomy and computers, even if she never always understood what I was up to. Thanks for keeping my focus on the stars.

Jeff Foust

1

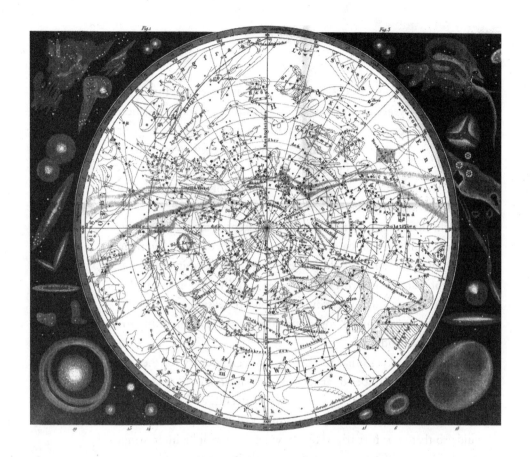

A GUIDED TOUR

THE SIZE OF THE UNIVERSE is measured in light-years, and no one knows its size for certain. The best estimates for the distance in light-years to the most distant objects known range from a few to more than 10 billion light-years. That's a pretty big number. In fact, it seems almost as big as the universe of software and online resources available on astronomy.

Well, there aren't quite that many different programs and Web sites available for you to choose from, but if you're just getting started the wide range of choices can seem a little overwhelming. What software should you use? Where should you look online for more information on an astronomy topic? It's easy to get lost and confused in cyberspace if you don't know what you're looking for or how to find and use it.

The situation is similar to visiting a foreign country for the first time. The sheer number of attractions—historic sites to see, museums to visit, restaurants to try, shows to attend, and so on—can overwhelm the new visitor. Sometimes you take a tour or hire a guide so that your first trip into a new territory will be interesting, delightful, and different, rather than simply incomprehensible.

With so many computer resources available for astronomical pursuits, it can be difficult to know where to go and what to do first if you're interested in a particular topic. This chapter will guide you through a hypothetical situation and introduce you to some of the computer resources available to answer your questions and guide you to more information if you're a little unsure about how to get started using your computer in astronomy, follow along on our guided tour.

A Quick Tour

It's a clear weeknight in late May, just after the Memorial Day holiday in the near future. After a long day at work and running errands, you don't get home until after nine in the evening. Just before you step inside, you look up into the southern sky and an unusual sight grabs your attention. You see the Moon and two bright stars below it. One of them has a definite reddish tint and is hardly twinkling. The juxtaposition of the three objects makes you curious. You know the brightest and largest is the Moon, but what about those other two stars? Why haven't you noticed them before?

Your home computer can help you answer those questions. Since you're wondering about the names and locations of objects in the night sky, the first place to turn to is a planetarium program you've already purchased and installed on your computer. You start up the program and start searching. Depending on what software you're using, you may have to input the time and your present location. For example, in some programs, like Starry Night Deluxe for the Macintosh and Windows (see Figure 1.1), you choose from a list of cities. You can save a default choice, however, so you won't have to do this each time you start the program.

Once you have the present location and time set, you turn to the southern horizon. Sure enough, there's the Moon, and there are the two bright stars below it. To find out what they are, you move

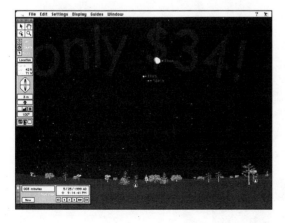

FIGURE 1.1
The Moon, Mars, and Spica as seen in Starry Night Deluxe

the mouse cursor over them and click on them. One is an ordinary, if bright, star: Spica. But the other is far more interesting: the "star" with the reddish tint is actually the planet Mars!

The motion of the Moon and Mars across the sky is why you didn't notice the two stars near the Moon before. Since the Moon is in a different part of the sky each night, it wasn't very near Spica and Mars last night, and it will be farther away tomorrow night. You can confirm this by adjusting the time settings in the program; when you travel backward and forward in time, you can see the distance between Spica and Mars change as Mars slowly moves across the sky in its own orbit. The next time the Moon is near Spica, Mars is much farther away. The close combination of the Moon, Mars, and Spica is a brief, ephemeral event that won't happen again for years or decades. (This is also something you can check in your planetarium program.)

Mars's appearance has piqued your curiosity. You know it's more than just a dot of light in the night sky: it's a planet, smaller than Earth but still large enough to have a thin atmosphere, giant mountains, and maybe, at one time, life. Your planetarium program might let you zoom in to get a close look at the planet, as if you were looking through a succession of more and more powerful telescopes. Some programs even let you look at maps of the surface, such as a close-up of the giant canyon Valles Marineris that stretches 3,000 miles across the planet (see Figure 1.2).

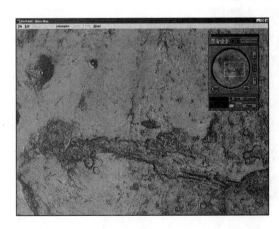

FIGURE 1.2
A map of Valles Marineris as seen in Redshift 2

There are limits to the planetarium program, of course. After using it to watch Mars move across the sky you want to find out more about the Red Planet than the program can provide. At this point you might turn to another software package with more information about Mars, as well as images of the planet taken by telescopes and spacecraft. One example is the Images of Mars program produced by the Exploration in Education program at the Space Telescope Science Institute for the Macintosh and Windows, available at **http://www.stsci.edu/exined/**. (You may already have a copy because the institute provides this and other programs free.)

Images of Mars provides a set of images of the planet taken by the Viking spacecraft, which landed on the planet in the mid-1970s. A set of 20 images offers views of the entire planet and detailed pictures of the surface, with captions that explain what you're seeing. Images of Mars can help you learn about topics ranging from Mars's giant canyons to the rocks strewn across the surface of the landing sites. Another program that provides a unique perspective on Mars is Mars Explorer from RomTech, available for Macintosh and Windows. This program gives you a 3-D perspective on the Martian surface, using data returned by NASA spacecraft to provide an accurate representation of the landscape (see Figure 1.3).

There is a limit, though, to the information you can get from software packages. They're static—only as up to date as when

FIGURE 1.3
A view of the Martian surface
generated by Mars Explorer

they're published. But astronomy is dynamic; new discoveries are made all the time, which means that the information software provides can quickly become dated. There's a solution to this problem—online, particularly on the World Wide Web, where continually updated information is only a click away, anytime you want it. You can even find updates for astronomy software on the Web.

So you fire up your Web browser—but you have to figure out where to go to find the latest about Mars. You could use a resource such as Yahoo, or a search engine such as AltaVista or Infoseek, but since you've read this book, you have a better idea. You start with a site offering general information about Mars—in this case, one that's part of The Nine Planets Web site (**http://www.seds.org/ nineplanets/nineplanets/mars.html**). Here you find some great images of Mars and some introductory information about the planet (see Figure 1.4). You could also find much of the same information at the Views of the Solar System Web site (**http:// spaceart.com/solar/eng/mars.htm**).

Scanning the page, you see some information about the discovery of possible evidence that primitive life existed on Mars billions of years ago. You remember hearing about this on the news when the announcement was first made, but you haven't heard much about it recently. A link to a NASA Web site (**http://rsd.gsfc.**

FIGURE 1.4
An introduction to Mars from The Nine Planets Web site

nasa.gov/marslife/) offers more information about the discovery, including some background on how scientists made the discovery and some images of the meteorite where they found the evidence.

Still curious about the discovery of life on Mars and what it might mean for the future of Mars exploration, you backtrack to the general Mars page and try another link, this time to a NASA site called the Center for Mars Exploration (**http://cmex-www.arc.nasa.gov/**; see Figure 1.5). There's some information about the Mars discovery here too, but mostly it's about plans for future missions to Mars, as well as current exploration, like two recent spacecraft, Mars Pathfinder and Mars Global Surveyor. You can follow links here to the Web sites for these and future Mars exploration spacecraft (**http://mars.jpl.nasa.gov**) to learn more about their missions.

A link near the top of the Mars Exploration page, titled The Whole Mars Catalog, beckons (see Figure 1.6). You visit this site (**http://www.reston.com/astro/mars/catalog.html**) and find a whole universe of links to Mars-related sites—the motherlode. If this collection can't satisfy your curiosity there's likely nothing online that will. (In fact, you could spend all night visiting these sites, and if they're interesting enough, you might do just that.)

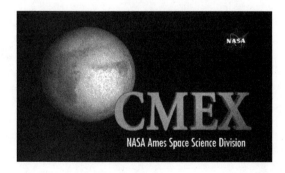

FIGURE 1.5
The Center for Mars Exploration Web site

The search we've described is just one example of how your computer could be an important, vital tool for understanding and exploring the universe. These times are unlike any other in human history—even the casual observer has immediate access to information, images, and other data about the universe and its phenomena, much of it as it's happening. To make sense of it all, though, you need to be able to find the information you really want. Let this book be your guide to the computer universe as you study and explore the real universe.

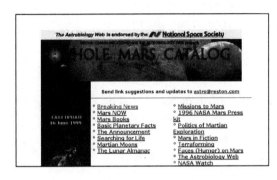

FIGURE 1.6
The Whole Mars Catalog has hundreds of links to Mars-related sites on the Web

2

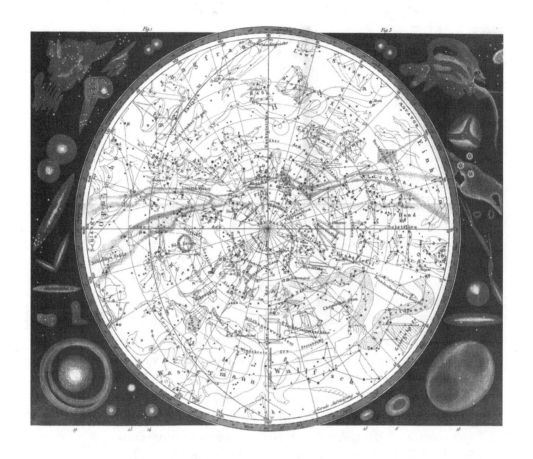

THE EARTH FROM SPACE

O N FEBRUARY 20, 1962, JOHN GLENN became the first American to orbit Earth. In his three-orbit spaceflight he used a simple 35mm camera, purchased for $45 shortly before the mission, to take photos of Earth. His three rolls of pictures were among the first images of Earth taken from space.

Since then astronauts have taken more than a quarter million pictures of Earth from orbit. Orbiting spacecraft have taken millions more images, providing information on Earth's weather, oceans, landforms, even structures beneath its surface. These images, and other information returned by satellites, have provided vital information that have allowed scientists to make more accurate forecasts, study the ozone hole in the Antarctic, watch for changes in the environment caused by human activities, and more.

These images of Earth from space have also shaped our view of the planet. While Earth seems very sturdy—literally, rock solid—to those of us on its surface, astronauts and cosmonauts have reported that the planet appears very fragile from space. One image of Earth—a view of Africa, Antarctica, and swirling cloud formations taken by Apollo 17 astronauts on their way to the Moon—has become one of the most popular pictures ever, appearing in a wide variety of places and becoming one of the environmental movement's symbols of Earth's beauty and fragility.

Many of these images, and other information, are now available to anyone using a computer. Some images come in software packages; many are accessible on the World Wide Web. This chapter looks at the wide range of images of Earth available to home computer users, with or without Internet access.

What Does the Weather Look Like?

For more than three decades, weather satellites have been sending detailed information, usually as images, to ground stations around the world. Many of these satellites are in geostationary orbits 23,000 miles (37,000 kilometers) above Earth, where it takes one day to complete an orbit. These satellites appear to hover over one region of Earth, returning a constant stream of data about the weather.

The primary form of data the satellites return is images showing the Earth's cloud cover. These images are of special interest to astronomers since they show how likely it is that the night will remain clear, or whether a cloud cover will break in time so the astronomers can make some observations. By following a series of satellite images taken over time, anyone can become an amateur meteorologist and guess when and where it will be cloudy.

Sources for Satellite Images

One of the best sources for satellite images is a Web site at the University of Wisconsin, **http://www.ssec.wisc.edu/data/**. This site provides satellite images for eastern and western North America from the GOES series of satellites, as well as global composite images assembled from data from several sources (see Figure 2.1). The two primary types of satellite photos available are visible

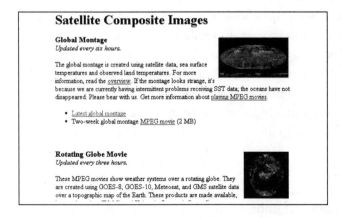

FIGURE 2.1
An online collection of current weather maps

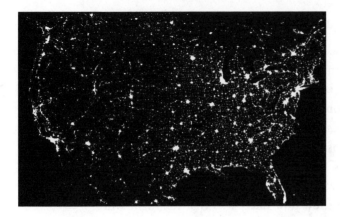

FIGURE 2.2
The United States at night, from a Defense Department weather satellite

wavelength, which provide the most detail during the day, and infrared wavelength, which return cloud images at night when visible wavelength images would be blank. Several movies, composed of a sequence of satellite photos, show how cloud patterns are changing over time. This site provides just the satellite images; analysis and interpretation are left up to you.

The Defense Meteorological Satellite Program (DMSP) collects another type of image via satellites in polar orbits (as opposed to the equatorial orbits of geostationary weather satellites) at much lower altitudes. While they can't provide the continuous coverage that geostationary weather satellites can, DMSP satellites provide more-detailed images of selected global regions on a daily basis.

The DMSP program maintains a site at **http://web.ngdc. noaa.gov/dmsp/dmsp.html**. While these images aren't immediately made available to the public (this is, after all, a military program), you can look at selected past images. One of the most interesting is shown in Figure 2.2—a picture of the United States on a cloud-free night. To an astronomer who wants dark skies for observing, this image is probably a little disheartening, even disturbing, as nearly every corner of the country is lit up. But it's a fascinating picture.

Tell Me about the Weather

If you'd rather let a professional tell you what the satellite photo means, there are a number of online sources of information about the weather. The Weather Channel has a significant presence on the Web at **http://www.weather.com**, with constantly updated conditions and forecasts for cities around the United States and in other countries. Similarly, Intellicast, a service of MSNBC, has forecasts, radar images, and satellite photos online at **http://www.intellicast.com**. The Weather Underground, at **http://www.wunderground.com**, provides a fancy interface for the otherwise dull forecasts and conditions the National Weather Service issues. There are many other sources of online weather information, particularly satellite images, but these sites are a good starting place.

Satellites That Monitor the Ozone Layer

Satellites also provide critical information on another important part of the atmosphere, the ozone layer—monitoring the seasonal hole that forms every spring in the ozone layer over Antarctica. Contrary to popular belief, satellites did not discover the ozone hole; scientists making measurements from the ground discovered the annual depletion of ozone high in the stratosphere. Later analysis of satellite data taken before the discovery confirmed the hole's existence, but this data had been ignored in the past, since the analysts thought any ozone readings that low must be in error.

NASA has mounted instruments in American, Russian, and Japanese spacecraft to monitor the ozone hole. These instruments, called *Total Ozone Mapping Spectrometers* (TOMS), measure the amount of ozone in a column of air from the surface to the upper atmosphere, although most of the ozone is concentrated in a single layer high in the stratosphere. TOMS make one measurement of the ozone content each day over nearly every portion of the planet.

NASA's Web site at **http://jwocky.gsfc.nasa.gov** provides information about its research into the ozone layer and the ozone hole. Visitors can view images like the one in Figure 2.3, which shows the ozone hole forming over Antarctica in October 1996. Whether natural processes or human activities cause the holes, this site provides a sobering look at the health of our atmosphere.

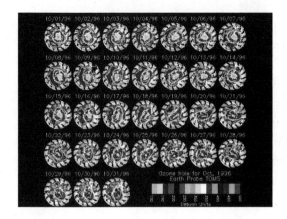

FIGURE 2.3
Images of the ozone hole over
Antarctica in October 1996

Satellite Pictures of Land and Sea

Satellites do more than provide information on the weather and
the Earth's atmosphere. A whole fleet of spacecraft takes images
and returns other data on the continents, oceans, islands, and bodies of water that make up the Earth's surface.

A number of spacecraft perform *remote sensing* activities. They
take images of the Earth's surface at various wavelengths; separately
or combined with other observations, these images provide information on the land, vegetation, human modifications, and more.
Government agencies and private companies prize this information, and use it to make decisions on land use.

LANDSAT and SPOT Images Can Be Very Pricey

The two biggest sources of remote sensing data are two sets of
satellites, LANDSAT and SPOT. LANDSAT was originally a
NASA project that has since been privatized. Information about
the LANDSAT project is available on the Web at **http://
edcwww.cr.usgs.gov**. Since LANDSAT images are sold to companies and the government, you have to pay for an image, sometimes
large sums of money. You can find a basic introduction to LAND-

SAT and its images, geared towards middle-school students, on the Web at **http://www.exploratorium.edu/learning_studio/landsat/ index.html.**

SPOT Image, a French company, has its own remote sensing satellites that compete directly with LANDSAT's. Information about its work is available on the Web at **http://www.spot.com.** Like LANDSAT, it sells its images for as much as thousands of dollars apiece. However, SPOT does provide a way to browse through its collection of 4.5 million images, and showcases special images, like a view of Oklahoma City after it was hit by tornadoes in the spring of 1999, as shown in Figure 2.4.

Some Sources for Free Images

Although these companies are the big commercial players in the field, they're not the only source. Not surprisingly, NASA has a very large collection of images of Earth from space shuttle

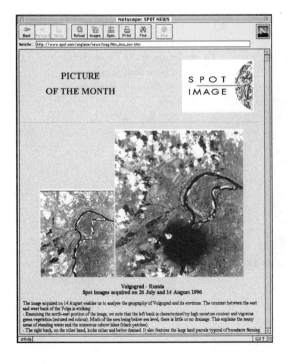

FIGURE 2.4
An image of Oklahoma City after a major tornado, taken by a SPOT satellite and displayed on SPOT's Web site

missions. Shuttle astronauts take images of Earth, either of prese-
lected regions or of whatever strikes their fancy, when they are not
busy working on their regular jobs during a mission. They have
amassed over a quarter million images of the planet.

A subset of this collection, about 30,000 images in all, is available at Earthrise from the San Diego Supercomputing Center at
http://earthrise.sdsc.edu. The site lets you search for images of
Earth based on a particular region (such as your home state) or on a
specific geological formation. A search of the Earthrise database for
images of Massachusetts turned up over 120 different images, such
as the image of the city of Boston and Boston Harbor shown in
Figure 2.5.

A similar selection of Earth images in Europe, a collection of
AVHRR images, is at **http://shark1.esrin.esa.it.** As with Earthrise,
you can search this database for images of a particular region of the
globe and come up with many different possibilities. You'll have to

FIGURE 2.5
**An image of the Boston area
taken from shuttle mission
STS-53, from the Earthrise
Web site**

pick through them carefully, since the database returns all images of the region, including those where clouds completely obscure it.

You don't have to be online to view great images of Earth. A number of software packages include collections of Earth images with descriptive information. One example is the application Endeavour Views the Earth, published by the Excellence in Education project at the Space Telescope Science Institute. This package includes dozens of images taken in 1992 by the space shuttle Endeavour on its maiden mission, complete with information about what's visible, even pointing out specific features in the photographs, as shown in Figure 2.6. More information about this free software package (Macintosh and Windows) is at **http://marvel. stsci.edu/exined-html/endeavour.html**.

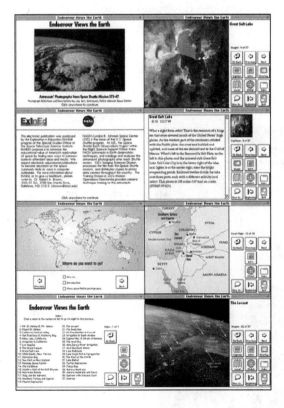

FIGURE 2.6
Screenshots from the program
Endeavour Views the Earth

There are also some low-cost commercial options. Colorado-based ARC Science Simulations provides very high-quality, high-resolution simulated images of the Earth. Its Web site, at **http://www.arcinc.com/**, includes information on its Face of the Earth Global Image Project. The site includes some samples to show how the quality of satellite images varies based on the angle of the Sun and atmospheric conditions (Figure 2.7), as well as samples of the company's own simulated images, including those for sale.

Pictures of the Earth's Oceans

Satellites do more than return images of the land—they can also show us the Earth's oceans from space. The TOPEX/Poseidon spacecraft, a joint project of the United States and France, has returned information on the state of the world's oceans since its launch in 1992. Its highly accurate altimeter allows it to measure the level of the seas and study their waves. Among the results returned from this mission is the discovery of a small rise in the global sea

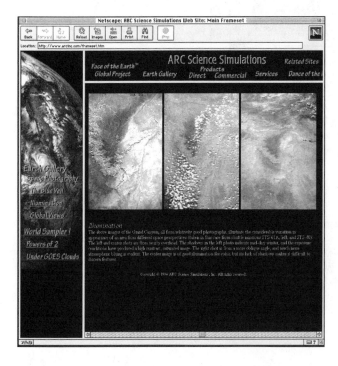

FIGURE 2.7
Three views of the Grand Canyon from orbit, with widely varying results, from the ARC Science Simulations Web site

level; not enough to cause any damage, but enough to raise concerns about future increases created by global warming. More information about TOPEX/Poseidon is online at **http://topex-www.jpl.nasa.gov**.

And Beneath the Surface

Some views from space allow us to look at layers below Earth's surface. Several shuttle missions have flown with a special radar mounted in the cargo bay that bounces radar off the planet's surface and uses the returned signal to construct images of the surface. These images provide information on layers immediately below the surface, since radar waves can penetrate a small distance. Information about these missions is online at **http://www.jpl.nasa.gov/radar/sircxsar/**.

One of the most amazing uses of these radar images was to help archeologists discover the Lost City of Ubar in Arabia. The city was a major trading outpost from about 2800 B.C.E. until A.D. 300, but all traces of it were lost long ago. Using radar images from shuttle missions, archeologists discovered trails that had once been used to travel to and from Ubar and found streambeds that had long since dried out (Figure 2.8). Taking this data into the field, the archeologists finally discovered the Lost City, and are working today on the excavation of this historic site.

Getting the Big Picture—the Earth in Space

While examining Earth from space can yield information on the weather, oceans, land types, and even ancient civilizations, perhaps the most important benefit of observations from orbit is the ability to place Earth, our home, in a new perspective.

The hundreds of cosmonauts and astronauts who have flown in space have said pretty much the same thing: They regard the Earth in a completely different way after seeing it from space. From space, they say, political boundaries aren't visible, allowing them to see and appreciate Earth in a new way. Apollo 9 astronaut Russell Schweickart said, "When you go around the Earth in an hour and a half, you begin to recognize that your identity is with that whole

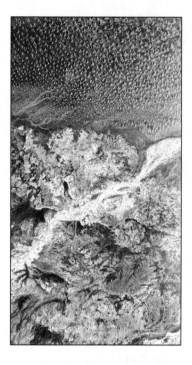

FIGURE 2.8
A radar image of the area around the
Lost City of Ubar

thing . . . you can't imagine how many borders and boundaries you cross, again and again, and you don't even see them." Author Frank White has termed this the "Overview Effect." You can read some of these comments from astronauts and cosmonauts on the Web at **http://spaceart.com/solar/eng/earthsp.htm**.

You can get some feel for this effect with your computer. One Web site, Earth Viewer (at **http://www.fourmilab.ch/earthview/ vplanet.html**), provides detailed images of the world as seen at any time from almost any vantage point, either from deep in space or orbiting just above the surface. You can add to the realism by including clouds and using recent images from weather satellites. A sample image is shown in Figure 2.9. Since the site is in Switzerland, it can be a little slow at times, but the images you will get out of it will be worth the wait. While it isn't the same as seeing Earth from space, it's a good online substitute.

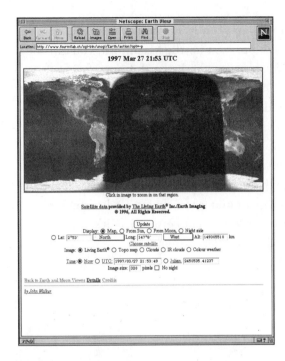

FIGURE 2.9
A sample from the Earth
Viewer Web site

FIGURE 2.10
A simulated view of the
northeastern United States
from Redshift

Software That Simulates Earth Views

Many planetarium programs can simulate views of Earth from space. Some allow you to adjust your altitude and rise from just above the ground to far away from Earth. For example, in Starry Night, a planetarium program for the Macintosh, you can increase your altitude and look down on the Earth from a few feet or a few million miles. Other planetarium programs provide a close-up of the Earth. Redshift 2, for Macintosh and Windows, generates simulated views of the Earth up to a moderate resolution: under 100 kilometers (62 miles) per centimeter on the screen (Figure 2.10). The program also lets you add the positions of major cities and observatories to the maps.

Now that we've used the computer to help learn a little more about the Earth and place it in a different perspective, it's time to move on to the other worlds in our solar system.

3

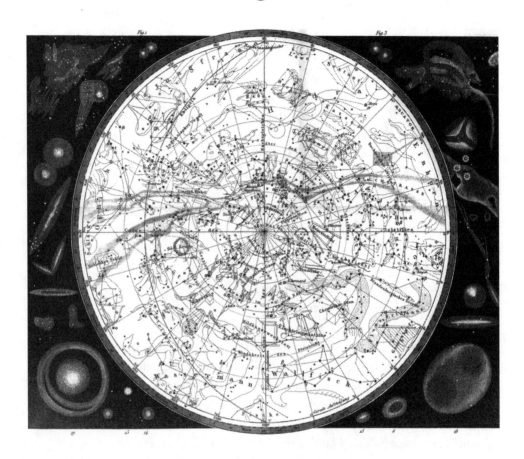

THE MOON

O F ALL THE OBJECTS IN the sky other than the Sun, the Moon probably has the greatest effect on Earth. When full, the Moon is the second-brightest object in the sky, and while it is much dimmer than the Sun, it is enough to brighten the night sky and make it easy to move around at night. The regular pattern of the Moon's phases and positions has influenced many animal species. The tides that the Moon generates may have played a crucial role in the evolution of life from water to land by washing organisms born in the oceans onshore and then stranding them on dry land. Throughout human history the Moon has played a role in military campaigns and has been immortalized in song and verse.

The proximity of the Moon to Earth has made it a prime target for space exploration in recent decades. The United States and the former Soviet Union launched dozens of unmanned spacecraft to the Moon in the 1960s and 1970s, ranging from simple spacecraft that returned a few pictures of the Moon as they flew by, to more ambitious spacecraft that landed on the Moon and automatically returned soil samples to Earth. Of course, there are the six Apollo moon landing missions, in which 12 astronauts became the only people that have set foot on a body other than Earth. Many of the information and results of these expeditions have been documented and made available to people with computers.

A wealth of information, images, and other resources about the Moon is available to computer users. This information ranges from accounts of lunar exploration to predicting lunar eclipses to thousands of images, available through the Web and through software packages. This chapter will provide a tour of some of the best online and offline resources about the Moon.

Learning about the Moon

It should be no surprise that the Moon is the most-studied body in the night sky. It has been the subject of naked-eye observations since antiquity and was one of the first objects Galileo studied with his telescope. More recently, it has been the target of a fleet of spacecraft from the United States and the former Soviet Union, and it is the only world other than Earth where humans have set foot. This intense analysis means we have learned a lot about our neighboring world, although many questions remain unanswered.

There is a wide array of information about the Moon available online. Two good starting places are the Moon entries in The Nine Planets (**http://seds.lpl.arizona. edu/nineplanets/nineplanets/luna.html**) and Views of the Solar System (**http://spaceart.com/solar/eng/moon.htm**). Both have general information about the Moon, including theories on its origin, its surface features, and histories of exploration. Both sites include many images and links too the online resources. The Royal Greenwich Observatory in England also has a brief summary of our knowledge of the Moon online at **http:// www.ast.cam.ac. uk/pubinfo/leaflets/solar_system/Moon.html**.

Educators looking for information about the Moon for their classes can turn to Spacelink, a site at NASA's Marshall Space Flight Center. This site at **http://spacelink. msfc.nasa.gov/Instructional.Materials/Curriculum.Support/ Space.Science/Our.Solar.System/Earth's.Moon/.index.html** includes information from the Apollo moon missions. It has a very plain interface, so you'll have to do a little exploring to find the information you're looking for, but it's worth a visit. There's also a good description of how the Moon creates tides in the ocean, useful for the classroom and the general public, by astronomer Phil Plait at **http://www.badastronomy.com/bad/misc/tides.html**.

Observing the Moon through your telescope is fascinating, providing an opportunity to observe first hand a world utterly different from our own. The light gathering power of telescopes, however, often makes the image of the Moon so bright that observing it can prove painful. This is a prime example where filters can come to your aid. Neutral density filters are colorless filters designed to simply cut down the amount of light passing through them, and are available in several "strengths," depending upon your needs. Such filters are usually threaded and typically screw into the back of an eyepiece. Some observers prefer a variable polarizing filters which, while more expensive, have the advantage of being able to vary the amount of light that is passed through them. Whichever is your choice, filters can provide a comfortable way to observe our nearest celestial neighbor. —R.L.

The Moon, like other bodies in the solar system, has a detailed and often complex system of nomenclature: the ways lunar features such as craters and mountains are named. For example, large craters on the Moon are named after famous deceased scientists, but small craters are given common first names. The U.S. Geological Survey has a thorough list of names and descriptions of lunar features compiled by its astrogeology branch, available at **http://wwwflag. wr.usgs.gov/USGSFlag/Space/nomen/moon/moonTOC.html**. You can view some of these features on the Web with John Walker's Moon Viewer at **http://www.fourmilab.ch/earthview/vplanet.html**. This site allows you to choose the vantage point from which to view the Moon: from Earth, the Sun, or another spot in the Solar System (see Figure 3.1). You can zoom in on features on the Moon's surface by clicking on them in the browser window.

Much of the imagery available in the Moon Viewer comes not from Earth-based observations but from the spacecraft reconnaissance

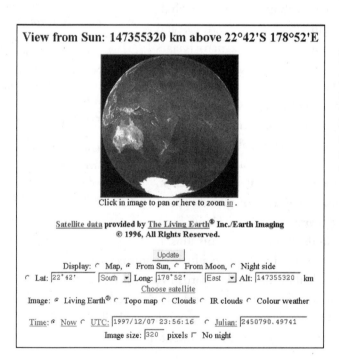

FIGURE 3.1
This image from the Moon Viewer shows the far side of the Moon, which is never visible from Earth; when the far side is completely illuminated, the near side is totally dark—a new moon

of the Moon in the 1960s and early 1970s. As the United States and the former Soviet Union raced to be the first to put people on the moon, they sent an array of unmanned spacecraft that returned valuable information on the Moon in preparation for manned missions. These spacecraft took the first images of the Moon's far side and provided details on the composition and features of the Moon's surface. It's no surprise, then, that much of the information available online discusses these missions. NASA's Johnson Space Center has an excellent overview of the American and Soviet missions in Exploring the Moon at **http://cass.jsc.nasa.gov/pub/expmoon/lunar_missions.html**. This site goes through both American and Soviet unmanned missions, such as Ranger, Surveyor, Lunar Orbiter, Luna, and Zond (which was probably a prototype for a Soviet manned mission to the Moon that never flew), as well as the Apollo Moon landings. This site has information and images about each type of mission.

The National Air and Space Museum has put together a detailed site on just the Apollo program at **http://www.nasm.edu/APOLLO/**. There is information on every Apollo mission to the Moon, with lots of images. There are also links to relevant resources available elsewhere online, such as the Apollo Lunar Surface Journal, a complete transcript of the astronauts' words and activities as they explored the lunar surface on the six manned lunar landings. If you need an offline source of information about the Apollo missions, the Space Telescope Science Institute's Excellence in Education project has developed Apollo 11 at Twenty-Five, a program for Macs and Windows that you can download from **http://marvel.stsci.edu/exined-html/Apollo.html** (see Figure 3.2). Created for Apollo 11's 25th anniversary in 1994, this program has information on the entire mission from liftoff to splashdown, illustrated with images.

Even simple binoculars show a surprising amount of detail on the lunar surface. If you're thinking about getting a pair of binoculars to use for astronomical purposes, there are several things you might consider. The novice usually thinks that the available magnification is the most important consideration when buying either a telescope or a good set of binoculars. In fact, the width of the lenses, which allow them to gather more light and thus see fainter objects, is more important than magnification. As magnification increases, binoculars become more and more difficult to steady, requiring a rigid mount of some sort by the time they get up to about 20x. A wider, less magnified field, usually provides more spectacular views. Combine this with the brighter image afforded by a wider lens, and views of the night sky can be spectacular, and the binoculars will be comfortable to use for extended periods of time.

High magnifications are infrequently used even on astronomical telescopes, except during the rare instances of great atmospheric clarity and then only on a limited number of celestial objects. When buying a telescope or a pair of binoculars, don't be misled by advertising claims touting high magnification figures.—R.L.

FIGURE 3.2
A screenshot from the
Apollo 11 at Twenty-Five
program

The Apollo program returned hundreds of pounds of lunar rock and soil samples for study by scientists. The Johnson Space Center has put information about the soil and rock sample program on the Web at **http://exploration.jsc.nasa.gov/curator/lunar/lunar.htm**. Included is information on how you can request lunar samples, either for research or classroom education projects.

Since the Apollo program there has been little exploration of or interest in the Moon. That trend has been changing in recent years, though, and more recent information about the Moon is becoming available online. The Galileo spacecraft, launched in 1989 on a six-year journey to its ultimate destination, Jupiter, passed by Earth and the Moon in 1990 and 1992, using Earth's gravity to pick up enough velocity to reach the giant planet. During each flyby, space-craft controllers at the Jet Propulsion Laboratory used Galileo's cameras and other instruments to record data on Earth and the Moon. Among the data Galileo returned on the Moon was more information about the Mare Orientale (see Figure 3.3), a giant basin that straddles the Moon's near and far sides that was likely created by a tremendous impact billions of years ago, and the South Pole Aitken Basin, a huge crater at the Moon's south pole. Informa-tion on Galileo's discoveries about the Moon is available at **http://www.jpl.nasa.gov/galileo/messenger/oldmess/Moon.html.**

FIGURE 3.3
The Galileo spacecraft took this image of the Mare Orientale, the bull's-eye in the center of the image. Mare Orientale sits partly on the near side of the Moon and partly on the far side, so an image like this of the full basin is only available from a spacecraft

In 1994 the Defense Department launched Clementine, a small spacecraft that went into orbit around the Moon for two months. Intended to test sensors and other equipment that could be used in ballistic missile defense satellites, Clementine returned detailed images and topological data about the Moon. Clementine also bounced radio waves off the Moon's south pole, including the Aitken Basin, in search of areas of permanent shadow where water ice might be found. In late 1996, scientists confirmed that there do appear to be deposits of water ice in these shadowed craters, which never see the light of the Sun and remain very cold. The Clementine maintains an updated Web site with information and images from the spacecraft at **http://www.nrl.navy.mil/clementine/clementine.html**.

In 1998, NASA launched the Lunar Prospector spacecraft, which surveyed Mars from a low orbit. While the spacecraft didn't carry a camera, it did carry spectrometers and magnetometers to survey the compostion of the Moon's surface and its magnetic fields. Within two months of entering orbit, it confirmed that large deposits of water ice appear to be hidden in the Moon's polar regions—a valuable resource for future human missions to the Moon. The project's Web site, **http://lunar.arc.nasa.gov**, has

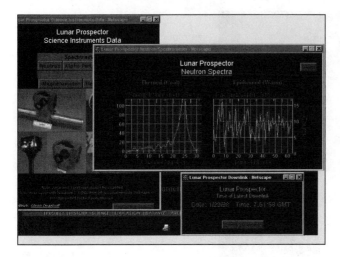

FIGURE 3.4
The Lunar Prospector Web site allows users to see the latest data returned by the spacecraft, such as data from the spacecraft's neutron spectrometers, displayed above

information about the spacecraft and the science of the mission, including the ability to see "real-time" data returned by the spacecraft (Figure 3.4).

Images of the Moon

Given the nearness and brightness of the Moon, coupled with the large number of spacecraft missions to it, you would rightly expect that there would be a lot of images of the Moon available online. Indeed, there are a number of excellent images on Web sites, most notably from the numerous manned and unmanned spacecraft missions.

Both The Nine Planets (**http://seds.lpl.arizona.edu/nineplanets/ nineplanets/luna.html**) and Views of the Solar System (**http:// spaceart.com/solarsys/eng/moon.htm**) have collections of Moon images taken from Earth and from spacecraft as part of the materials about the Moon. These sites, especially The Nine Planets, serve as a good starting point in the search for images.

The Regional Planetary Image Facility, at the National Air and Space Museum, has a large collection of images of the Moon at **http://www.nasm.edu/ceps/RPIF/MOON/moon.html**. This

FIGURE 3.5
An image of the Earth rising above the lunar surface, as seen from the Apollo 11 spacecraft in orbit around the Moon; one of hundreds of such images available from the Johnson Space Center

collection of images includes many images taken by Apollo and unmanned spacecraft during their missions to the Moon. For an extensive collection of images from the Apollo program, visit **http://cass.jsc.nasa.gov/pub/expmoon/apollo_landings.html**. There are links here to NASA's large collection of public images from all of the Apollo missions, including many good images of the Moon and Earth from space (see Figure 3.5). The National Space Science Data Center (NSSDC) has a collection of Moon images, including many from spacecraft, on its Web site at **http://nssdc.gsfc.nasa.gov/ photo_gallery/photogallery-moon.html**.

Some of the sites mentioned above also have good collections of Moon images. The National Air and Space Museum's Apollo information (**http://www.nasm.edu/APOLLO/**) includes many images of the Moon taken during the Apollo missions. The Exploring the Moon site (**http://cass.jsc.nasa.gov/pub/expmoon/lunar_missions. html**) has images of the Moon and the manned and unmanned spacecraft that traveled there. The Clementine Web site (**http://www.nrl.navy.mil/clementine/clementine.html**) also has dozens of images taken from its more recent mission. In short, there are plenty of images of the Moon available online or in software (such as the Apollo 11 at Twenty-Five site mentioned above) to keep any Web surfer busy for a long time.

Observing the Moon and Its Effects

While the Moon may be one of the most fascinating objects for planetary scientists and those interested in space exploration, for many astronomers the Moon is nothing short of Public Enemy Number One. Since the Moon is the second-brightest object visible from Earth, its appearance in the night sky makes it difficult to observe fainter objects lost in its glare. In fact, the time around the full moon, when the Moon is at its brightest and is up all night long, is usually not used by astronomers except for those looking at bright objects. This time is often devoted at major observatories to routine maintenance of the telescope and facilities.

Thus, for astronomers, knowing the phase of the Moon is nearly as important as knowing the weather. If the Moon is full, it makes little difference whether the skies are clear or cloudy. The phase also indicates when the Moon will be visible. A full moon rises at sunset and sets at sunrise, while a new moon, which occurs about two weeks before and after each full moon, rises at sunrise and sets at sunset. The first-quarter Moon (which occurs between the new and full moon, when half the Moon's face is illuminated as seen from Earth) is at its highest point in the sky at sunset and sets by midnight, while a last-quarter Moon (midway between the full moon and the new moon when again half the Moon's face is illuminated) rises around midnight and is at its highest point in the sky by sunrise. If this seems a little confusing, you may want to consult The Educator's Guide to Moon Phases, which describes Moon phases for teachers and is available at **http://spaceart.com/solar/eng/edu/moonphas.htm**.

Essentially any planetarium program provides information on the Moon's phase and its rise and set times. Some programs, like Distant Suns for Macs and Windows, can provide a month long calendar of Moon phases with a single mouse click (see Figure 3.6).

Shareware programs, such as MOONPHILE (Macs, **http://rtd. com/~gschneid/MOONCLOCK.html**) and moontoolw (Windows, **http://www.fourmilab.ch/moontoolw/**), provide information on the current state of the Moon, including its exact phase, location in the sky, and more. Some Web sites can also provide this and other information about the Moon's phase. Lunar Outreach Services can give you the exact phase, down to the percent of the Moon's face that is illuminated, at **http://www.lunaroutreach.org/**. This site includes a rendition of how the moon looks in its current phase. Images of the moon's past phases are also available here. You can get something similar from the Moon Viewer mentioned earlier (**http://www.fourmilab.ch/earthview/vplanet.html**), with more detailed images and the ability to change the date for which it computes and displays the phase of the Moon.

For those interested in planning their observations a month at a time, the Phases of the Moon Web site at **http://www.googol.com/ moon/** will be useful. This site has calendars that show the Moon's phase for each night over the course of a month. This makes

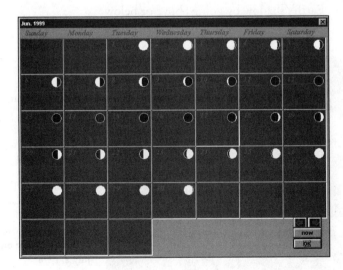

FIGURE 3.6
A calendar of Moon phases and other astronomical events generated by Distant Suns 4.0 for Windows

33

planning a month's observations to avoid the full moon (or to observe it) very easy.

If you're looking for information on the phase of the Moon at different times throughout history, visit A View of the Moon at **http://saatel.it/users/lore/moon.html.** This site can generate the phase of the moon for each night over the course of 20,000 years, from 9999 B.C. through A.D. 9999 (see Figure 3.7).

One lunar phenomenon that's of interest to all astronomers, including those who dislike the Moon because of its brightness, is a lunar eclipse. During a lunar eclipse, the Sun, Earth, and Moon line up so that the Moon passes through Earth's shadow. Since Earth is much larger, the Moon can spend several hours in its shadow; the normally bright full moon turns into a dim, ruddy orb. No wonder many astronomers like lunar eclipses: For a few hours, the effect of moonlight on the visibility of other objects in the night sky is greatly diminished.

You might expect that lunar eclipses should take place every month, since once a month the Moon lies behind Earth as seen from the Sun. However, since the Moon's orbit is slightly inclined, most of the time it misses Earth's shadow by going above or below

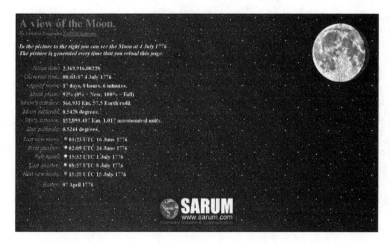

FIGURE 3.7
A screenshot from the View of the Moon Web site shows that the Moon was just a few days past full when the American colonists signed the Declaration of Independence on July 4, 1776

it, so no eclipse takes place. Details behind solar eclipses are explained in the Educator's Guide to Eclipses, available online at **http://spaceart.com/solar/eng/edu/eclipses.htm**. (This site also explains solar eclipses, when the Moon comes between Earth and the Sun.) More technical information is at **http://www-astro. physics.uiowa.edu/~jdf/lec24/node1.html**, which are the online notes about eclipses and eclipse predictions from an astronomy class at the University of Iowa.

You could try to find lunar eclipses with a planetarium program, scanning through the dates to find a time when the Moon is partially or totally within Earth's shadow (see Figure 3.8). However, NASA has compiled a complete online list of lunar eclipses for a thousand-year period. Visit **http://sunearth.gsfc.nasa.gov/eclipse/ LEcat/LEcatalog.html** for a list of lunar eclipses between the years 1501 and 2500. This information is part of a larger site, **http:// sunearth.gsfc.nasa.gov/eclipse/eclipse.html**, with general information about solar and lunar eclipses.

There is also software to simulate and predict lunar eclipses. An eclipse simulator program for PCs is available for downloading at **http://www.astronomy.ch/html/lunareclipse1.htm**. This shareware

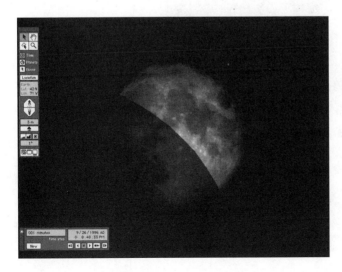

FIGURE 3.8
A simulation of a total lunar eclipse seen from North America on the night of September 26, 1996, using the planetarium program Starry Night Basic 2.0

program simulates the circumstances of solar and lunar eclipses over a several-hundred-year period. A free Mac program, UMBRAPHILE, can be downloaded from **http://rtd.com/ ~gschneid/UMBRAPHILE.html**. This bare-bones program provides the predicted times of an eclipse, given certain information about the eclipse and the observer's location. You can also use this program to run a camera, automating the process of taking pictures during an eclipse. Some planetarium programs, such as The Sky, by Software Bisque, provide predictions of lunar and solar eclipses and show what the eclipse will look like from a given area on the planet (see Figure 3.9).

There is other information about lunar eclipses online as well. Anthony Mallama's lunar eclipse page, at **www.serve.com/meteors/ mallama/lunarecl/index.html**, has information about the times of lunar eclipses and how to observe them. Australia's Calwell Lunar Observatory has a site at **http://www-clients.spirit.net.au/~ minnah/LEO.html**; it includes the predicted time of the eclipses and efforts of network observers throughout the world to observe them. Finally, a set of pictures of a typical lunar eclipse, taken by a

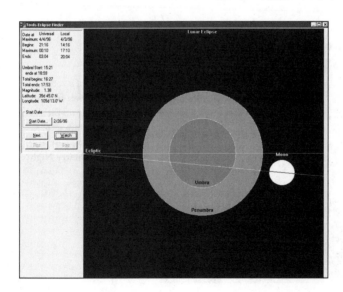

FIGURE 3.9
A schematic of the April 3, 1996, total lunar eclipse as seen from the Northeastern United States

Spanish scientists, is available for people to view at **http://www.geocities.com/CapeCanaveral/5993/**. These images show an eclipse from the beginning stages to its maximum extent, when the face of the Moon has turned a dark reddish brown (caused by the Sun's light refracting through Earth's atmosphere and reaching the Moon.)

There's more information about the Moon available, much of it accessible from sites listed in this chapter. The resources listed here, both online and offline, will provide a good start for anyone looking for more information about the Moon, for research, general knowledge or to plan observations. In the next chapter we turn from the Moon to the other planets of the solar system and their moons.

4

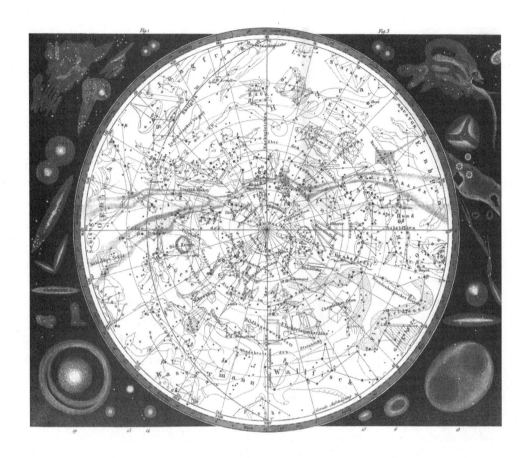

PLANETS AND MOONS

O UR KNOWLEDGE OF THE solar system has exploded in the last 40 years. We're no longer limited to the tiny, fuzzy images available from telescopes on Earth; we have sent dozens of spacecraft from Mercury to the far reaches of the solar system and beyond, visiting every planet but Pluto. These spacecraft have returned a wealth of information about these very different worlds, including countless amazing, dramatic images of planets and moons far more different and unusual that we could have imagined.

As it turns out, these missions have raised at least as many questions as they have answered. Based on images of the surface, where landforms have been clearly shaped by water, we know that water probably once flowed on Mars, but we don't know what happened to end that warm, wet period. Voyager 2 images of Neptune showed a large, dramatic storm on the planet, which has disappeared in later images taken with the Hubble Space Telescope. But we don't know why Neptune's atmosphere is so active while Uranus, about the same size as Neptune and much closer to the Sun, shows far less activity. There is still a lot to learn about the planets of the solar system and their moons.

This chapter will explore some of the online and software resources available for discovering what we do know about the solar system. Given the dynamic nature of this research, with new images returned literally every day by spacecraft such as Galileo (which is in orbit around Jupiter), we emphasize online resources that that are frequently updated with new information and images. We'll still look at software programs for learning more about the solar system offline.

Basic Resources

The limited scope of the solar system—there are only nine planets after all, including the Earth—makes it fairly easy to summarize our knowledge of each world. A number of Web sites and software packages do this, usually with some of the thousands of images of the planets spacecraft have returned in the past few decades.

Two of the best online resources about the solar system are Views of the Solar System and The Nine Planets. Both sites feature detailed information about each planet, as well as their moons, with lots of updated news and images. In particular, Views of the Solar System features many images specially analyzed by its creator, Calvin Hamilton, while The Nine Planets includes thorough historical notes written by site creator Bill Arnett, with information about the mythological origin of the planets' names. There are also links to related sources on the Web. You can access Views of the Solar System at **http://spaceart.com/solar/**; The Nine Planets is online at **http://www.seds.org/nineplanets/nineplanets/**. Both sites have mirrors on other computers around the world for easier access, especially for users outside North America.

There are several less comprehensive general resources about the solar system online. The Jet Propulsion Laboratory, which has served as mission control for a number of planetary spacecraft missions, including Viking, Voyager, and Galileo, features Welcome to the Planets on its Web site. This site, at **http://pds.jpl.nasa.gov/planets/**, includes many images of planets, mostly taken by NASA spacecraft. Another JPL site, the Planetary Photojournal, **http://photojournal.jpl.nasa.gov**, allows users to search a database of thousands of images of planets, moons, and other solar system bodies taken by NASA spacecraft over the years. The information on the site, though, is limited to the basics. The National Space Science Data Center, a repository of data collected by spacecraft missions, has a similar feature in its NSSDC Photo Gallery at **http://nssdc.gsfc.nasa.gov/photo_gallery/**. This site also features some fact sheets with data about each planet at **http://nssdc.gsfc. nasa.gov/planetary/planetfact.html**. Rounding out the bottom of the list are some smaller collections of images, including Arizona State University's DSN Solar System Menu at **http://esther.la. asu.edu/asu_tes/TES_Editor/dsn_solarsyst.html**, and the Regional Planetary Image Facility, a part of the Center for Earth and Planetary Studies at the Smithsonian's National Air and Space Museum, which is available at **http://www.nasm.edu/ceps/rpif.html**.

While there are a number of good collections of planetary information online, plenty of information is available in software packages that don't require online access. The disadvantage of these is that they cannot be updated easily with new information, an important factor given the amount of new data rolling in from telescopes and spacecraft. But they can include many large images and animations that would be too unwieldy on the Web, especially for people with older, slower modems. One example is The Planetary System, a multimedia tour of the solar system produced by the Space Telescope Science Institute for Macintosh and Windows computers. This site features information and images on planets and moons in the solar system in an easy-to-use format that would be more difficult to implement on the Web. However, the package doesn't have updated information on recent discoveries, like the possibility of past life on Mars and the latest news from the Galileo spacecraft orbiting Jupiter. The software is free, though, so you have nothing to lose but a few minutes of downloading time. It's accessible on the Web at **http://www.stsci.edu/exined/ Planetary.html**. A companion program, PlanetQuest, discusses the spacecraft missions that have revolutionized our knowledge of the planets; it's available at **http://www.stsci.edu/exined/ planetquest.html**.

The Discovery Channel Interactive's Beyond Planet Earth provides a multimedia exploration of the planets for Windows users. This program combines images and videos that look at our exploration of the solar system to date and plans for future missions.

There are also interviews with experts such as former astronauts Buzz Aldrin and Kathryn Sullivan.

Star Trek fans will probably enjoy The Nine Worlds (Mac and Windows), hosted by Patrick Stewart. He provides dramatic narration for extended videos about each of the nine planets, plus the Sun. There is also more specific information about the planets and human exploration of them, from telescopes to spacecraft. Users looking for more will appreciate the links to online information.

If you're looking for something more interactive than narrated videos on your computer, check out Solar System Explorer from Maris, for Macs and Windows. In this program, which is more like a game than an educational program, you are working on a space station early in the next century (see Figure 4.1). You not only learn more about the planets, you can design and launch missions to them!

Observing these distant planets can prevent a challenge for the amateur astronomer. Under some magnification, Uranus and Neptune can be distinguished as small globes in amateur telescopes, or at least will appear distinctly different from nearby stars. Pluto, however, will only appear as a star. Locating Pluto requires a medium-to-large amateur telescope and a fairly detailed star chart of the area. Even then, it is often only realized after careful observations spanning several sessions that one particularly faint starlike object has moved against the background stars.—R.L.

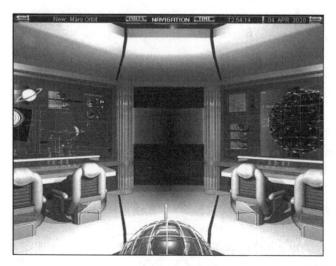

FIGURE 4.1

The Navigation Center in Solar System Explorer lets you plan missions to the other planets

The Inner Planets: Mercury and Venus

Despite their closeness to Earth, the two inner planets of Mercury and Venus are still among the most mysterious in the solar system, for a number of reasons. Since both are closer to the Sun than the Earth is, they always appear in the sky near the Sun, and thus can't be seen except for a few hours before sunrise or after sunset, depending on their orbital positions relative to Earth. This is especially troublesome in the case of Mercury, which is the planet closest to the Sun and thus never strays very far from it in the sky, making it the hardest-to-see of the bright planets. Venus, on the other hand, appears for longer times in the night sky from Earth, but until this century, and most notably this past decade, we have learned little about what exists under the continuous thick layer of clouds that hides its surface from view.

However, our state of ignorance about Venus is slowly changing, thanks in large part to a single spacecraft mission in the early 1990s. The Magellan spacecraft, which launched in May 1989 and arrived at Venus a little over a year later, used radar to penetrate the clouds and provide a global, high-resolution map of the planet's surface. The images revealed a surface considerably different from our own, with unusual craters; large, extinct volcanoes; and no evidence of the plate tectonics that shape and resurface our world (creating earthquakes and volcanic eruptions in the process.) The National Space Science Data Center maintains a good introduction to the Magellan spacecraft, including images of it and links to other resources, at **http://nssdc.gsfc.nasa.gov/planetary/magellan.html**.

Magellan returned literally gigabytes of data on the surface, more than all previous NASA planetary missions combined. Scientists have toiled to turn these virtual mountains of raw data into useful results, several of which are available online. NASA's Planetary Data System has put much of the data online; find out how to access this at **http://delcano.mit.edu/http/midr-help.html**. However, it helps to know which of NASA's dozens of CD-ROMs has the images you want, something the average person will have no information about. Another, easier to use resource is the Venus

Hypermap, http://www.ess.ucla.edu/hypermap/Vmap/top.html, a clickable map of the planet created by the Earth and Space Science Department at UCLA that displays its color-coded topography. You can click on an area to zoom in on smaller regions and see more detail.

Even more sophisticated is the Face of Venus project, http://stoner.eps.mcgill.ca/~bud/craters/FaceOfVenus.html, among the earliest (created in 1994) and most useful examples of an interactive Web site. It features a full database of all the craters that Magellan recorded on the planet's surface. You can search for certain types and/or sizes of craters on the planet, and get a dynamically generated map with the crater locations identified. The site also includes a database of coronae—large, unusual circular features on Venus's surface that don't appear anywhere else in the solar system.

One of the most comprehensive maps of the planet is available in a CD-ROM package called Venus Explorer by RomTech. Available for Macs and PCs, it includes a high-resolution map of the surface of Venus collected from Magellan data. (There are some gaps where data was lost due to problems with the Magellan spacecraft or with communications between Magellan and Earth.) You can zoom in on regions of the map to get better-resolution images (Figure 4.2), and also search for specific features by name. However, there's not much written information or documentation to go with this program, so unless you know just what you want, or are content with pretty pictures, you may be left puzzling out what exactly you were just looking at.

If a couple hundred megabytes of poorly labeled images don't interest you, the Space Telescope Science Institute's Magellan Highlights of Venus package, available free for Macs and Windows at http://www.stsci.edu/exined/Venus.html, may be a better solution. This collection of images of Venus taken by Magellan, with detailed descriptions, is better for people who are less interested in global coverage and more interested in learning about the unusual features of Venus's surface (Figure 4.3).

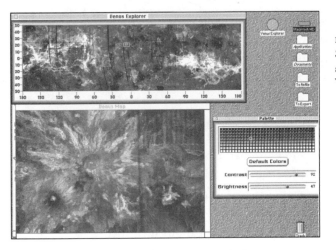

FIGURE 4.2
Studying a region of Venus's surface in the program Venus Explorer

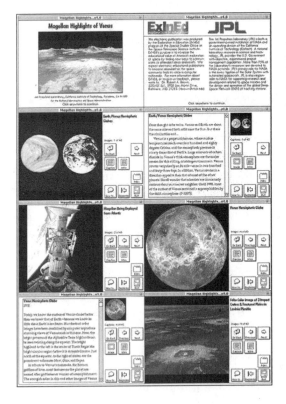

FIGURE 4.3
Screenshots from the Magellan Highlights of Venus program showing radar images of the planet and what we've learned about the planet from them

While Magellan and previous spacecraft missions are helping us understand our twin planet, we are far less enlightened about its inner neighbor, Mercury. Its location close to the Sun in the sky makes it difficult to observe, and makes it hard to get a spacecraft there without a lot of fuel and proper shielding from the Sun's radiation. Only one spacecraft, Mariner 10 in the mid-1970s, has returned images of Mercury, which appears in many areas to resemble the heavily cratered regions of our own Moon. However, scientists using radar waves, bounced off the planet back to Earth, have shown that there may be deposits of ice in shadowed regions near the planet's poles, regions that may never see the blinding light of the Sun. News about this possible discovery is online at http://nssdc.gsfc.nasa.gov/planetary/ice/ice_mercury.html.

Mars

Mars has captured the attention of humans perhaps more than any other object in the night sky, save the Moon; the vast number of online resources and software about the Red Planet reflect this interest. Its striking red appearance led many ancient peoples to associate it with war and destruction. Later, as Mars came under the scrutiny of telescopes, it emerged as a world of its own. Changing surface features and hotly debated observations of "canals" on the face of Mars led some to believe that not only was Mars inhabited, it was home to a race of intelligent beings, struggling to find water on an increasingly dry world. Naturally, some portrayed these hypothetical Martians as malevolent, sending them on fictional invasions of the Earth to claim our own more hospitable world, as in H. G. Wells's *War of the Worlds*.

It's ironic, then, that the fictional Martians of Wells were undone by mundane terrestrial microbes and bacteria, since it appears that the real Martians might once have been nothing more than microbes themselves. The August 1996 announcement that a meteorite traced back to Mars contained evidence of biological activity sent the astronomical world into a frenzy of discussion and

47

debate, and rekindled our interest in the Red Planet. While the debate over whether the meteorite offers any real evidence of past life is unresolved, interest in Mars remains high.

Within literally hours of the 1996 announcement about the discovery of possible past life on Mars, Web sites appeared with more information and links to other resources. NASA provides the official scoop on the discovery on its Evidence of Primitive Life from Mars site at **http://rsd.gsfc.nasa.gov/marslife/**. This site includes the press releases announcing the discovery, background information, and images and animations of the meteorite and the evidence of life. JPL also has a page of general information on Mars meteorites—known as SNC (pronounced *snick*) meteorites, for the three types of Mars meteorites, Shergotty, Chassigny, and Nakhla—at **http://www.jpl.nasa.gov/snc/**.

Several other comprehensive Mars life resources are available. John Pike of the Federation of American Scientists' Space Policy Project created its Mars life site at **http://www.fas.org/mars/ index.html** just a few hours after the announcement, and has kept it up-to-date with additional information and links to other sites. Reston Communications has created the Whole Mars Catalog; as the name suggests, this is a comprehensive collection of information about Mars, with an emphasis on the possibility that life may exist somewhere on Mars today. This resource is online at **http://www.reston.com/astro/mars/catalog.html**.

Interest in life on Mars certainly predates the discovery of possible biological evidence in a Martian meteorite, though. In 1976 the Viking mission landed two spacecraft on Mars, equipped with instruments that could detect microbes or other living creatures in the Martian soil. After some initial confusion over the results, scientists decided that the experiments showed that the soil was very active chemically, but contained no evidence of life. Malin Space Science Systems, a company that builds instruments for planetary spacecraft missions, including those to Mars, provides a good summary of these Viking experiments at **http://barsoom.msss.com/ http/ps/life/life.html**. General information on the Viking missions

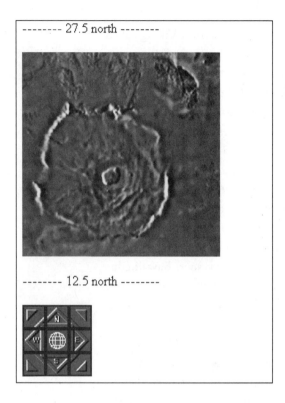

-------- 27.5 north --------

-------- 12.5 north --------

FIGURE 4.4
An image of the giant extinct volcano
Olympus Mons on Mars, as seen by the
Viking spacecraft and displayed in the Mars
Atlas at the NASA Ames Research Center

to Mars is available from the NSSDC at **http://nssdc.gsfc.nasa.gov/
planetary/viking.html.**

Although the Viking missions did not find evidence of life on
Mars, they did give us a detailed global picture of the planet,
thanks to two spacecraft that remained in orbit while the other two
landed on the surface. The two orbiters took thousands of images
of the surface and returned data on its composition and on the thin
Martian atmosphere. Several Web sites have made versions of these
images available for people to browse. The Mars Atlas at NASA's
Ames Research Center allows you to select and zoom in on a region
of the Martian surface, using Viking images. The site doesn't give
you information on what you're looking at, but if you know a specific feature you want to examine, like the giant volcano Olympus
Mons (Figure 4.4), this site can meet your needs. It's available on
the Web at **http://ic-www.arc.nasa.gov/ic/projects/bayes-group/**

49

Atlas/Mars/. A similar database of Mars images from Viking called Mars Explorer for the Armchair Astronaut has a similar format, although its global image of Mars also provides you with the names of key features on the planet. It can be accessed at **http://www-pdsimage.wr.usgs.gov/PDS/public/mapmaker/mapmkr.htm**.

There's more one can do with the images returned by Viking than just display it in flat images. A group at the University of Hawaii has taken images of the Martian surface around the landing sites for the two Viking spacecraft and put them into an interactive QuickTime VR movie. The movie provides the illusion of a third dimension, and you can turn around and zoom in and out to get a view of what the Viking spacecraft saw when it landed on Mars. The site is available on the Web at **http://www.soest.hawaii.edu/PRPDC/mep/qtvr.html** and requires the QuickTime VR plug-in for Netscape or Internet Explorer to view the landscapes in the browser window.

An extension of this idea is available in a CD-ROM package called Mars Rover. This program by RomTech converts Viking images of Mars into landscapes for a number of predefined regions on the planet. You can take guided tours and scroll around, looking at the scenery in detail (Figure 4.5). The program is available for Macs and PCs on the same two-disc set.

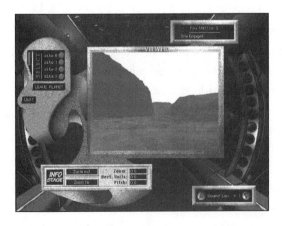

FIGURE 4.5
A three-dimensional view of the Martian landscape from the Mars Rover program

FIGURE 4.6
An image of the Ares Vallis region of Mars taken by Mars Pathfinder

Fascination with the Red Planet grew in 1997 with the successful landing of the Mars Pathfinder spacecraft. For several months, Pathfinder and its small rover, Sojourner, took images and returned data on a region of Mars that was likely a site of a catastrophic flood in a warmer, wetter period of Mars's history. Although the mission ended many months ago, the Mars Pathfinder Web site is still available at **http://mars.jpl.nasa.gov/default.html** and has archived information about the mission, including many high-resolution images (Figure 4.6).

While Mars Pathfinder's mission has long since ended, another Mars mission is still in progress. Mars Global Surveyor, launched around the same time as Mars Pathfinder, has been returning images and other data from its orbit around Mars since September 1997. The project's Web site, **http://mars.jpl.nasa.gov/mgs/index.html**, provides status reports on the mission and links to images returned by the spacecraft, including images that show that the so-called "Face on Mars" seen in Viking photos is just an ordinary, rather un-facelike rock formation (Figure 4.7).

Two more spacecraft are scheduled to arrive at Mars by the end of 1999. Mars Climate Orbiter, launched in December 1998, will go into orbit around Mars in September and study the planet's climate and weather systems. Mars Polar Lander, launched in January 1999,

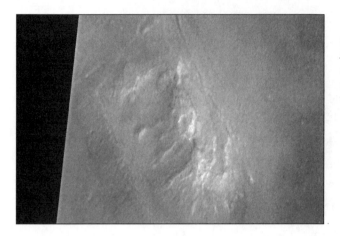

FIGURE 4.7
A Mars Global Surveyor image of the infamous "Face on Mars" shows it looks far less like a face than first thought

will land near Mars's south polar cap in December. It will look for evidence of water ice hidden in the planet's soil, among other things. It also carries two small spacecraft, part of a separate mission called Deep Space 2, that will separate from the lander prior to landing and crash-land on the Martian surface, burying probes up to two meters deep into the soil to study the composition of the subsurface soil. Information on these missions, as well as planned missions in 2001 and beyond, is available at **http://mars.jpl.nasa.gov.**

Not surprisingly, the news of possible past life on Mars has sparked interest in future exploration of the Red Planet, by unmanned spacecraft as well as humans. A good clearinghouse for information about the potential and the realities of human and robotic exploration of Mars is the Center for Mars Exploration at NASA's Ames Research Center, **http://cmex-www.arc.nasa.gov/.** This site has links to current and planned unmanned missions and possible future human exploration. One of those human missions is the Mars Direct mission. Designed by aerospace engineer and writer Robert Zubrin, Mars Direct would use Mars's own natural resources, such as the carbon dioxide in its atmosphere, to help produce rocket fuel for the trip home, making the spacecraft lighter and less expensive without compromising safety. Complete information on the mission, including some detailed technical papers, is on the

Web at **http://www.nw.net/mars/**. There are also links to other spacecraft and Mars science resources at **http://www.seds.org/~spider/mars/mars.html**. If you or someone you know is looking for reasons to be convinced that Mars exploration is a good idea, NASA engineer Michael Duke has written an essay on this topic at **http://spaceart.com/solar/marswhy.htm**, part of the Mars portion of the Views of the Solar System site.

The idea of sending people to Mars has already attracted the attention of artists, who have created a number of works portraying human exploration of the planet. The West to Mars site, at **http://www.marswest.org/**, has dozens of contributions by artists showing how people would live and work on Mars as they explored the planet, just as people explored and settled the western frontier of the United States 150 years ago. Mars has also long been the subject of scientific and science-fiction books. A detailed bibliography of these books, from classic texts to modern-day releases, is available at the Web site Mars in the Mind of Earth, **http://www-personal.engin.umich.edu/~cerebus/mars/**.

Jupiter and Saturn

Jupiter is probably the most influential planet in the solar system. As the largest planet, it has gravity strong enough to hold a miniature solar system of moons, including four that are larger than some of the planets. It has swept clean certain regions of the asteroid belt: Its gravitational resonances kick any asteroids that wander into this region—including those that cross Earth's orbit—into new orbits. It also influences the orbits of comets: for example, shortening the orbital period of the giant comet Hale-Bopp from over 4,000 years to under 2,500. Jupiter is king of the planets.

But the king is vulnerable to surprisingly small bodies. In July 1994, professional and amateur astronomers around the world, as

Even through a good pair of binoculars, Saturn presents an unusual sight – appearing to be shaped something like a football, rather than the expected spherical object. With a larger instrument, the oblong shape is obviously caused by the spectacular ring system for which this planet is known. While we now know that other planets such as Jupiter and Neptune have rings, none are nearly as bright, complex, or as easily observed as those of Saturn.

The disk of Saturn is much more subtly colored than that of Jupiter, and it is usually much more difficult to observe the banding and storms on the surface of the planet. A filter, particularly one with a bluish tint, can be a great aid to observing Saturn. Such filters are readily available, come in a variety of colors, and can be a valuable aid in observing planetary objects.—R.L.

well as members of the general public who rarely turn their eyes to the night sky, gazed in wonder as fragments of a small comet—probably no more than a mile or two in diameter—slammed into Jupiter at nearly 40 miles a second, triggering gigantic explosions and dark impact spots that lasted for weeks. Tiny comet Shoemaker-Levy 9, named for the team of astronomers who discovered the shattered comet in 1993, managed to bruise the mighty planet. It was one of the most violent cosmic events humans in this solar system have ever witnessed.

The impact of Shoemaker-Levy 9—or SL9, as it was more simply known—took place around the time the Web was gaining broad popularity, and thus became the first astronomical event to be widely publicized online. The Jet Propulsion Laboratory assembled a Web site to serve as an interface to a collection of nearly 1,500 images and animations of the comet impact as seen by dozens of observatories around the world. The Web site is still available at **http://www.jpl.nasa.gov/sl9/**. If the 1,500-image figure is too intimidating, the site maintainer has a list of the 20 most popular images, based on the number of times they were accessed. This list includes a number of images from ground-based telescopes and from the Hubble Space Telescope (Figure 4.8); it's online at **http://www.jpl.nasa.gov/sl9/top20.html**.

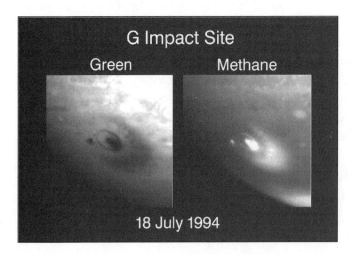

FIGURE 4.8
Hubble Space Telescope image, from the JPL SL9 Web site, showing the large, dark impact site created by fragment G, one of the largest fragments to hit the planet, in two different wavelengths

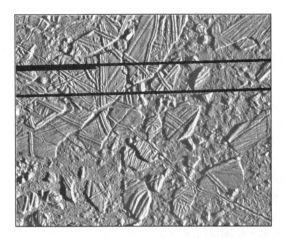

FIGURE 4.9
This image of Europa, taken by the Galileo spacecraft and available from the Galileo Web site at JPL, shows ice rafts that, like icebergs on Earth, may be pulled around the surface by currents from a liquid water ocean under the surface

There are a number of other Web sites with information and images of the comet collision. A close second to the JPL site is a site hosted by Students for the Exploration and Development of Space, a pro-space exploration student group. This site includes information from before and after the impacts, hundreds of images of the impacts from observatories around the world, and a comprehensive list of links to other SL9 Web sites, all at **http://www.seds.org/sl9/**. This site also has a simple browser to search for images from each of the approximately 20 impacts, at **http://www.seds.org/ sl9-impact/**. For some general information about the comet, including lists of frequently asked questions (FAQs), turn to **http://www.isc.tamu.edu/~astro/sl9.html**.

More recently, attention has turned to the new information about Jupiter and its large moons returned by the Galileo spacecraft, which went into orbit around Jupiter in December 1995. Although a faulty main antenna has hampered the spacecraft's ability to return some data, it has still managed to send back thousands of images of Jupiter and its four largest moons: Io, Europa, Ganymede, and Callisto. The images of Europa are perhaps the most interesting, since they show that the icy moon's crust may be only a mile or two thick, with a deep layer of liquid water underneath it (Figure 4.9). With heat from the moon's interior and traces of organic chemicals already discovered, all the ingredients are in place for primitive life to exist in Europa's subterranean oceans.

JPL has assembled a detailed Web site at **http://www.jpl.nasa.gov/galileo/** with complete information about Galileo and its mission, including new images of Jupiter and its moons, updated daily, as well as news releases and background information about the mission. Another part of the Galileo spacecraft was the probe, which plunged into Jupiter's dense atmosphere the same day the main spacecraft went into orbit around the planet. The spacecraft returned interesting information on the composition and structure of Jupiter's atmosphere that had not previously been considered. NASA's Ames Research Center ran the probe, and it has its own Web site at **http://ccf.arc.nasa.gov/galileo_probe/**. The Galileo mission is designed to follow up on the observations of Jupiter the two Voyager spacecraft made 15 years previously; information about the Voyager mission and images from the Jupiter system are online at **http://vraptor.jpl.nasa.gov/voyager/voyager.html**.

While Galileo studies Jupiter and its moons, a similar large spacecraft will soon explore the planet Saturn and its moons. The Cassini spacecraft, named after the Italian astronomer who investigated the rings of Saturn with a telescope several hundred years ago, launched in October 1997 and will arrive at Saturn in 2004. Like Galileo, it will carry an atmospheric probe—only this probe, named Huygens and built by the European Space Agency, will fly through the atmosphere of Saturn's largest moon, Titan, and perhaps survive a landing on Titan's surface, which like Venus's surface is shrouded from view by a dense layer of clouds and haze. JPL has assembled information about the mission, including the instruments the spacecraft will carry and the scientific questions it will try to answer, at **http://www.jpl.nasa.gov/cassini/**.

The two Voyager spacecraft have visited Saturn as well as Jupiter, and the Voyager Web site at **http://vraptor.jpl.nasa.gov/voyager/voyager.html** has information about the studies Voyager made at Saturn, along with images of the planet and its moons. Specific information about Saturn's complex ring system, based largely on observations from the Voyager spacecraft and some ground-based data, is online at **http://ringside.arc.nasa.gov/www/saturn/saturn.html**. If you just need some information on Saturn-related events, such as transit times (when Saturn's moons appear to cross the disk of the planet as seen from Earth), and when the planet will eclipse its moons, turn to **http://www.isc.tamu.edu/~astro/saturn.html**.

The Outer Planets: Uranus, Neptune, and Pluto

We know even less about the outer planets in the solar system than we do about Jupiter and Saturn, as these planets, especially Pluto, are so far from Earth they're harder to observe in telescopes and much more difficult to send spacecraft missions to. The Voyager Web site at JPL has the best information about Uranus and Neptune: turn to **http://vraptor.jpl.nasa.gov/voyager/vgrur_fs.html** for a summary of Voyager 2's study of Uranus and **http://vraptor.jpl.nasa.gov/voyager/vgrnep_fs.html** for a summary of Voyager 2's Neptune studies. Voyager images of Uranus and Neptune are also available on the Web site at **http://vraptor.jpl.nasa.gov/voyager/vgrur_img.html** for Uranus and **http://vraptor.jpl.nasa.gov/voyager/vgrnep_img.html** for Neptune.

Both Uranus and Neptune have ring systems, but they are different from the immense rings surrounding Saturn. Uranus has a series of nine major but thin rings, which were not discovered until 1977 when astronomers on an aircraft observing the planet passing in front of a star noticed that the star blinked several times as the planet moved past it. The existence of the rings surrounding Neptune wasn't confirmed until Voyager 2 flew by in 1989; its results showed unusual clumps of material in parts of the rings. Information about the ring systems of both worlds is online at **http://ringside.arc.nasa.gov/www/uranus/uranus.html** and **http://ringside.arc.nasa.gov/www/neptune/neptune.html**.

Ironically, Pluto, the smallest planet in the solar system and the one usually most distant from the Sun (its elliptical orbit brings it closer to the Sun than Neptune for about 20 years in each of its orbits around the Sun, which last over 250 years; it was be closer to the Sun than Neptune from 1979 until 1999), has more online resources dedicated to it than some other, less distant worlds. Much of this may be related to fascination many have with the small, icy, mysterious world, which was not discovered until 1930, when young astronomer Clyde Tombaugh noticed it in a pair of photographic plates of a section of the night sky he had taken earlier that year. The University of Colorado's Pluto Home Page has a good introduction to the planet, its history, and our current knowledge of it, on the Web at **http://dosxx.colorado.edu/plutohome.html**. You can learn more about the life of Clyde Tombaugh, who passed away in early 1997, at **http://www.klx.com/clyde/**.

There is a lot of research in progress right now to better understand the planet, including attempts to find basic information like its mass and composition. One of the key scientists in this research is Marc Buie at Lowell Observatory in Arizona, who has used the Hubble Space Telescope to produce some of the best images yet of what the surface of the distant world may look like (Figure 4.10). His Web site, at **http://www.lowell.edu/users/buie/pluto/pluto.html**, includes information about Pluto and his current research, and links to other Pluto sites.

Our current information about Pluto is limited to observations from the Hubble Space Telescope and telescopes on the ground; Pluto is the only planet in the solar system that spacecraft have not visited. Engineers and scientists at JPL and other institutions are working to change this with a proposal for a spacecraft mission called Pluto-Kuiper Express. This would include one or two small spacecraft (in case one failed in flight, and also to get close-up images of both sides of the planet, since it takes over six days to complete one turn on its axis), launched just after the turn of the century and arriving at the planet about 10 years later. The spacecraft would also fly by one of the objects in the Kuiper Belt (see below). Full information about the proposal, which has not yet been approved by NASA and is subject to the often-fickle funding

are seeing the side of Charon that faces away from Pluto. Get image

Sub-earth latitude = 12 degress, longitude = 95 degrees. On the left side of the Pluto image is the Charon-facing hemisphere. The bright streak at the south pole of Pluto is a bad spot in the map. This side of Pluto is much darker than the rest. At this view, Pluto is 30% darker than on its brightest side. Get image

FIGURE 4.10
Computer models of what the surface of Pluto may look like, based on images of the planet taken by the Hubble Space Telescope and posted on Marc Buie's Web site

moods of Congress, is available at **http://www.jpl.nasa.gov/ice_fire/pkexprss.htm.**

Pluto is by far the smallest planet in the solar system. In fact, it is so small that some astronomers have suggested it not be classified as a planet at all, but as an asteroid or the largest member of a class of small icy bodies in the outer solar system known as the Kuiper Belt. Although there are important historical reasons for considering Pluto to be a planet, it's true that Pluto does share some characteristics with asteroids and comets. The next chapter will look at these two types of small bodies in the solar system.

5

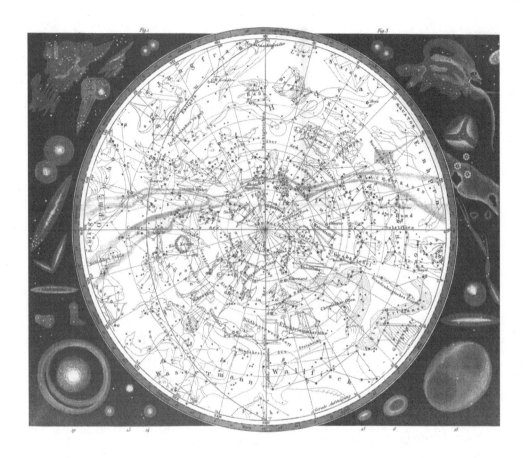

ASTEROIDS AND COMETS

A T ONE TIME, THE SOLAR SYSTEM seemed to be a very simple place. There was Earth, the Sun, the Moon, and several other planets. Even after Earth was displaced from the center of the Solar System (and the universe), and even after the discovery of moons around other planets, there seemed to be a clear order to the Solar System. The addition of two classes of objects changed all that. The discovery of asteroids at the beginning of the nineteenth century added a whole new class of bodies to the Solar System, and although comets had been watched since ancient times, they were not completely understood as a separate class of bodies in the Solar System until a few hundred years ago. The use of telescopes soon meant that even more comets not visible to the naked eye were added to the Solar System's inventory. Now, instead of consisting of a handful of planets and moons, the Solar System was populated with thousands, perhaps millions, of small rocky and icy bodies in a wide range of orbits. This chapter will look at these two classes of numerous bodies and the resources for learning more about them on the computer.

Asteroids and Meteors

On the night of January 1, 1801—the first night of the nineteenth century —Sicilian astronomer Giuseppi Piazzi made the first observation of what became the first known asteroid, Ceres. He did not make this observation was not made by chance; for years, astronomers had been scanning the skies for a missing planet about 2.2 times farther away from the Sun than the Earth is. The Titius-Bode Law, a mathematical relationship that calculated the distances of the planets from the Sun, predicted the existence of such a planet. It also predicted the location of a planet beyond Saturn, which was discovered in the 1780s and named Uranus. But it also called for a planet between Mars and Jupiter no astronomer had seen—hence the search, coordinated by a group of astronomers known as the "Celestial Police."

The discovery of Ceres, although it was disappointingly dim and small, seemed provide the solution. However, in March 1802, German astronomer Heinrich Olbers found another small body orbiting at around the same distance from the Sun as Ceres. This body was named Pallas. Soon more objects were found in similar orbits, and it became clear the "missing planet" was really a collection of small bodies, which became known as minor planets or asteroids.

Today we have well-known orbits for over 10,000 asteroids. Many more asteroids have been seen, but their orbits are not known; they may have been seen only once as a passing blur in an image an astronomer took of another object. Most of these asteroids are in the main asteroid belt between Mars and Jupiter, but many have been discovered in other orbits, including some that pass near Earth and even cross Earth's orbit. These bodies are of special interest and concern because of the threat they pose to Earth.

Not surprisingly, two of the best places to turn online for information about asteroids are The Nine Planets and Views of the Solar System. Each site has a good introduction to asteroids, with links to more information sources. Visit **http://www.seds.org/nineplanets/ nineplanets/asteroids.html** and **http://spaceart.com/solarsys/eng/ asteroid.htm** for more information about asteroids. The Royal Greenwich Observatory also provides some basic information, although not as much as The Nine Planets and Views of the Solar System; check out **http://www.ast.cam.ac.uk/pubinfo/leaflets/solar_ system/minor.html** for its offerings.

When asteroids were first discovered, they were given names from Greek and Roman mythology, in the same tradition as the planets' names. However, with over 7,000 asteroids now known and named, even the rich mythology of the Greeks and Romans falls short in providing enough names. Today, names of asteroids are given by their discoverers, and cleared by the International Astronomical Union, which has a few guidelines (no profane names, nor any names of living or recently deceased political leaders, for example). This opens up a huge possibility of names. Some asteroids are named after fellow astronomers to honor their work, but many are named after towns and other places, and others are named after people from arts, sports, and popular culture. To prove this point, visit **http://cfa-www.harvard.edu/cfa/ps/special/ RockAndRoll.html**, home to the "Rock and Roll Minor Planets" page, a list of asteroids named after rock and roll stars. You'll find that Eric Clapton, Jerry Garcia, Frank Zappa, and all four members of the Beatles have been immortalized.

Since asteroids are dim enough not to have been discovered until long after telescopes were widely used in astronomy, it requires some effort to observe one. Planetarium programs are a big help here, as most include a database of at least a set of the brightest asteroids, which would be the easiest for observers to see. Some programs, like Redshift 3 for Macintosh and Windows (Figure 5.1), provide good supplementary information about asteroids, including images, background, and detailed positional data.

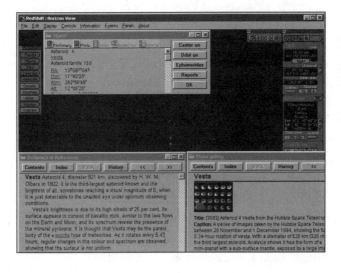

FIGURE 5.1
Information about the bright asteroid Vesta, provided by Redshift

If you're looking for more information about asteroid positions, particularly dimmer, more obscure asteroids, MPO99 for Windows 3.1 and 95 (as well as previous versions from previous years, such as MPO98 and MPO97) is an excellent piece of software. This program includes a database of over 7,000 asteroids, far larger than what a typical planetarium program offers. For any given night, MPO99 can tell you not only what asteroids are visible, but which ones are located near stars, galaxies, and other asteroids in the sky (see Figure 5.2). You can also create star charts showing the location of the asteroid relative to other stars, to make locating the asteroid much easier. This is a useful tool if you're hunting for dimmer asteroids or trying to identify a possible asteroid that appeared serendipitously in an image.

Many people who observe asteroids are interested in *occultations*, when the asteroid passes in front of a distant star. By timing how long the asteroid blocks the star's light, astronomers can estimate the asteroid's size. By combining observations of a single occultation made over a large area, astronomers can also get a better idea of the asteroid's shape and orbit. The International Occultation Timing Association's Asteroid Occultation section has a Web site at http://www.anomalies.com/iota/splash.htm with information about

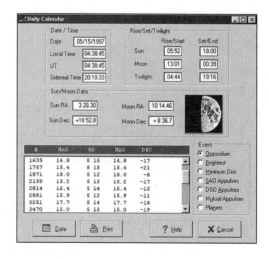

FIGURE 5.2
MPO97's Daily Calendar lists which asteroids are in opposition, meaning they will be visible all night long, as well as which asteroids will pass near bright stars and other objects

recent and upcoming asteroid occultations. The site also has information on Asteroid Pro (**http://www.anomalies.com/asteroid/info.htm**), software that amateur astronomers can use to predict and plan for asteroid occultations visible from their location.

There is a considerable amount of research available from ground-based telescopes that are designed to look for new asteroids and understand the nature and composition of known ones. A good starting place for current news and more information is the Minor Planet Center, part of the Harvard-Smithsonian Center for Astrophysics at **http://cfa-www.harvard.edu/~graff/mpc.html**. This site also has information about the Spaceguard Foundation, an

---

My love of the night skies goes back deep into my long-ago childhood. One of my earliest memories is returning home with my family from a trip to some long forgotten destination, riding on my uncle's shoulders and rapturously gazing up at what seemed to be the late Carl Sagan's "billions and billions" of stars blazing from the heavens. That sight was no less moving to my childhood eyes and imagination than Van Gogh's Starry, Starry Night is to my adult sensibilities. As best as I can determine, it was that night that I became an amateur astronomer – a pastime that endures to this day.

In school I absorbed everything astronomical that I could put my hands upon. I learned that falling stars are really only rocks and pebbles burning as they fall into Earth's atmosphere, that there was a planet with rings— Saturn, and I learned of a "once in a lifetime" celestial visitor–Comet Halley–that appeared in Earth's skies once every 76 years. For some reason, Comet Halley captivated my imagination and its next apparition became an event that I awaited patiently for some 30 years.

As an adult I learned that there were several planets that had ring sys-

tems, and I (along with millions of others) had a ringside seat to the flyby explorations of the outer planets of our solar system by robotic Voyagers who brought back answers to many of our questions and in turn gave us new questions to ask. But with my heart and mind I followed the reports that Comet Halley had been acquired by the largest of telescopes as an extremely faint blur as it slowly made its way from the outer fringes of the solar system to its appointed dance with the sun.

Over time the comet grew closer and could be seen in telescopes whose diameters measured in feet rather than yards,

then in inches rather than feet, and then scientists reported that it should soon be visible in the smaller telescopes of amateurs. Needless to say, I was out with my telescope every clear night, searching the region of the night sky in which its appearance was predicted. I searched many nights without success.

One such night however, just at the limit of visibility, I saw the faintest of blurs next to a small grouping of stars. This featureless blur appeared as nothing more than a faint nebula, although from my experience with telescopic viewing of faint comets, I thought this

effort to raise awareness about the dangers of near-Earth asteroids and the need to search for and catalog them, to know which ones pose the greatest threat.

One such effort is the Near-Earth Asteroid Tracking (NEAT) telescope, located on Haleakala, a mountain on the Hawaiian island of Maui. This relatively new facility uses advanced cameras and computers to discover thousands of asteroids, many which pass near Earth. The project's Web site at **http://huey.jpl.nasa.gov/ ~spravdo/neat.html** includes information about the project and the latest news, including images of recent discoveries.

---

might possibly turn out to be the long-awaited visitor. I left the telescope tracking this part of the sky while I searched and finally found the grouping of stars in my sky atlas. Yes, there was the star field I was observing, but there was no sign of any known nebulae, galaxy, or other deep space object adjacent to them. Back to the eyepiece for another look. And the faint glowing object had changed position! After 30 years of waiting, I was at last seeing the "once in a life-time" comet.

There have been few events in my life in which I found myself completely transported. There was a sense of

completion, a long quest successfully filled, and a sense of the great cycles of history sweeping me up. This faint object, in its present state tremendously unexciting to most observers, was perhaps the most exciting thing I had ever seen. Its history had changed empires, flooded humanity with terror at the approaching end of the world, and captured the attention of humanity like clockwork every 76 years. And now it was my turn, this was my moment with history, with Comet Halley.

Although this particular apparition of Comet Halley was not one of the best for Earth-based viewers due to the

positions of both the Earth and the comet, it nevertheless was a magnificent event. This apparition was unique in at least one particular – the famous comet was visited and observed by spacecraft. It is my understanding that with modern astronomical equipment that Comet Halley will never really be out of sight again, it can and will be tracked throughout the course of its orbit.

For most amateur astronomers Comet Halley has long since faded beyond the capabilities of their telescopes to observe. Someday in the next century, however, there will be reports of the great comets

approach and there will be those who follow such reports with a mixture of both curiosity and longing. They will no doubt begin to search the region of the night sky for the appointed return of the celestial visitor, and one night a faint blur that moves against the background of the stars will be seen. And the cycle will repeat.
—R.L.

Telescopes of all kinds are used in asteroid research. Scott Hudson, a researcher at Washington State University, uses radar to study asteroids, bouncing radio signals off them to get information about their shapes, which would not be visible using ordinary Earth-based telescopes. His Web site at **http://www.eecs.wsu.edu/ ~hudson/Research/Asteroids/index.html** describes the research he and his colleagues are working on, as well as offering some interesting links to other asteroid Web sites.

Despite their large numbers, few asteroids have been studied in detail by spacecraft. In the course of its winding six-year journey from Earth to Jupiter, the Galileo mission to Jupiter passed two main belt asteroids: Gaspra and Ida. This was the first opportunity to get a close-up look at an asteroid. The results revealed a number of surprises, including the discovery of a tiny moon, later named Dactyl, in orbit around Ida (Figure 5.3). This was the first, and to date the only, known case of an asteroid with its own moon. The Nine Planets and Views of the Solar System provide good information about both asteroids. Visit **http://www.seds.org/nineplanets/ nineplanets/gaspra.html** and **http://spaceart.com/solar/eng/ gaspra.htm** for information about Gaspra and **http://www.seds.org/ nineplanets/nineplanets/ida.html** and **http://spaceart.com/solar/ eng/ida.htm** for Ida and Dactyl information.

Another mission currently in progress will provide more information about asteroids. The Near-Earth Asteroid Rendezvous (NEAR) mission was launched in February 1996. In June 1997 it passed the main belt asteroid Mathilde, returning images of the asteroid. It was scheduled to arrive at the asteroid Eros in early 1999 and go into orbit around that small body, but a spacecraft glitch has delayed its arrival until February 2000. The spacecraft will return high-resolution images and other data on the asteroid for several months. This will be the first detailed reconnaissance of an asteroid and, if successful, will set the stage for future missions to learn more about the resource potential of asteroids. The NEAR team has an excellent Web site at **http://near.jhuapl.edu/** that includes a well-written educator's guide explaining the mission and what scientists

FIGURE 5.3
This image from Galileo shows
Ida, the large asteroid on the
left, and its small moon Dactyl,
on the right

FIGURE 5.4
The home page for the NEAR
Web site

hope to learn at a level students can understand (Figure 5.4). In addition, a provate company, SpaceDev, is planning its own asteroid mission. The Near-Earth Asteroid Prospector (NEAP), a small, low-cost spacecraft, will launch in 2001 to study the asteroid Nereus. More information on that mission is online at http://www.spacedev.com/NEAP/NEAP.html.

There is another way to get a close-up look at asteroids, without ever leaving Earth. Each year thousands of tons of tiny asteroids, far too small to be detected in space, encounter Earth. Most burn up in the atmosphere, but a few are large enough to reach Earth's surface. The ones that burn up in the atmosphere are the meteors or shooting stars you can see on occasion in the night sky. Those objects that survive the trip to the surface are meteorites. They provide scientists with samples of asteroid material that would cost far more to obtain from spacecraft missions.

Collecting meteorites has become a hobby for many and even a profession for a few. Hobbyists and serious collectors will appreciate the information available on two Web sites: the Meteorite Exchange (http://www.meteorite.com) and Meteorite Central (http://www.meteoritecentral.com). Both provide meteorite news and information; classified ads for people seeking to buy, sell, and exchange meteorites; and areas in which to talk with fellow collectors.

There's also some general information about meteorites at The Nine Planets, http://www.seds.org/nineplanets/nineplanets/meteorites.html, and Views of the Solar System, http://spaceart.com/solar/eng/meteor.htm. More information for observers is available from the International Meteor Organization at http://www.imo.net/.

Small asteroids make it to Earth's surface with regularity; larger asteroids hit the Earth less often, but with greater force; they are capable of doing catastrophic damage. Astronomers and others have become more aware of the threat of asteroid impacts in recent years as more small asteroids—only a few meters in diameter, but with enough power to level a city if they hit the Earth—have been seen passing near the Earth, in some cases closer to it than the Moon. Interest has also grown as more evidence appears that a large asteroid or a comet, hitting the Earth near the present-day location of Mexico's Yucatan Peninsula, triggered the mass extinction of the dinosaurs and many other species.

The Asteroid Comet and Impact Hazards Web site at the NASA Ames Research Center, http://impact.arc.nasa.gov/, provides a great deal of information about the threat and possible effects of asteroid and comet impacts (Figure 5.5). One of the sobering statistics available here is that the odds of dying from an asteroid impact are about 1 in 20,000, comparable to the odds of dying in a plane crash. While asteroid impacts are far rarer than plane crashes, they are also far deadlier. For another look at the role of asteroids and comets in extinctions, see Asteroids, Comets, Death and Extinction at **http://ourworld.compuserve.com/homepages/Ponder/doomsday.htm.** The National Geographic Society has created an interactive Web site, Asteroids: Deadly Impact, where users play the role of a detective trying to uncover evidence of asteroid impacts in the Earth's past from Mexico to Arizona to Russia (Figure 5.6). It's on the Web at **http://www.nationalgeographic.com/asteroids/.**

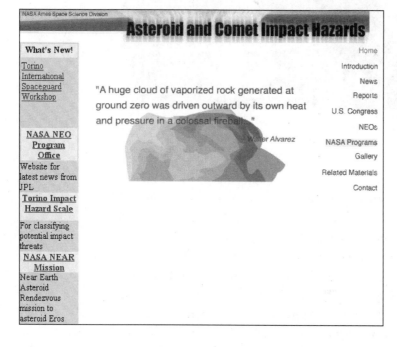

FIGURE 5.5
The Asteroid Comet and Impact Hazards Web site will let you know if the sky really is falling

FIGURE 5.6
Hunt for clues regarding impact-related events around the world with National Geographic's Asteroids: Deadly Impact Web site

Comets

Asteroids, like comets, are small bodies that likely are remnants from the formation of the Solar System four and a half billion years ago. The similarities all but end there, though. While asteroids are in orbits predominantly in the inner Solar System, comets are usually found on very elliptical orbits, taking decades, centuries, and even millennia to complete a single orbit yet spending only a few months of that time in the inner Solar System. Asteroids were not discovered until the beginning of the nineteenth century; comets have been known—and feared—since antiquity. Viewers need binoculars and telescopes to see asteroids; some of the brightest comets are easily visible even in the day.

The Nine Planets and Views of the Solar System both provide good introductions to comets: visit **http://www.seds.org/nineplanets/**

nineplanets/comets.html and http://spaceart.com/solar/eng/comet.
htm for the basics about comets. The Comet's Tale is an online education module for students in grades 4 through 12 created by
astronomers at the University of California at Berkeley (Figure 5.7),
at http://cse.ssl.berkeley.edu/comod/com.html. This site has information on comets, their history, and even how to make your own
"comet" using dry ice, water, and other household materials. The
Royal Greenwich Observatory also has introductory information about
comets at http://www.ast.cam.ac.uk/pubinfo/leaflets/solar_system/
section3.15.html. Comet expert Gary Kronk has assembled some historical information at his Web site, http://medicine.wustl.edu/
~kronkg/index.html.

Comets are believed to evolve into the inner Solar System from
the Oort Cloud, a shell of perhaps billions of cometary bodies
orbiting the Sun 50,000 to 100,000 astronomical units (AU) away.
By comparison, the Earth is only 1 AU from the Sun. It's believed
that when a star makes a relatively close approach to the Sun, it dis-
turbs the Oort Cloud, knocking some of the bodies into orbits that

FIGURE 5.7
The Comet's Tale Web site is an
educational exploration of comets

take them gradually into the inner Solar System, where the planets, particularly Jupiter, alter their orbits.

The bodies in the Oort Cloud may be linked to another source of cometary bodies, the Kuiper Belt. The Kuiper Belt is a disc of large cometlike bodies orbiting the Sun at Neptune's distance and beyond—in fact, it's been suggested that the planet Pluto and its large moon, Charon, may be the two largest bodies of the Kuiper Belt. Unlike the Oort Cloud, the Kuiper Belt is more than just a theoretical construct. Several dozen bodies up to a few hundred kilometers in diameter have been discovered in the last few years in the Kuiper Belt. For more information about the Kuiper Belt and the Oort Cloud, check out **http://www.seds.org/nineplanets/ nineplanets/kboc.html** and **http://spaceart.com/solar/eng/kuiper. htm**. We are also beginning to discover bodies in the distant Solar System that appear to be neither Kuiper Belt nor Oort Cloud objects, but something between. An example is 1996TL66, discovered in late 1996, which appears to be a large, icy object in a highly elliptical orbit 35 to 150 AU from the Sun. There's more information about this unusual object online at **http://cfa-www.harvard.edu/ cfa/ps/pressinfo/1996TL66.html**.

At any given time, there are no more than a few comets passing through the inner Solar System, and most of those are rather dim and hard to see. Unlike asteroids, many thousands of which have well-known orbits that can be plotted years in advance, comets are unpredictable. They come in orbits that bring them close to the Sun every few years to every few *million* years. Moreover, new comets are discovered at the rate of a couple dozen a year, and on occasion one of the new discoveries makes a brilliant appearance in the night sky, sometimes with only a few weeks' warning, like comet Hyakutake in 1996. Thus, it's hard to predict what comets, bright or dim, will appear in the night skies in upcoming months, let alone years.

You can get some idea of what is visible, though, by looking at the positions of known comets in a planetarium program. Most

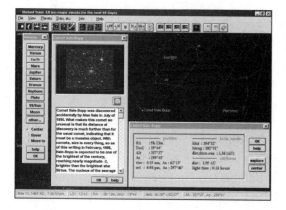

FIGURE 5.8
Comet Hale-Bopp as displayed by the
program Distant Suns

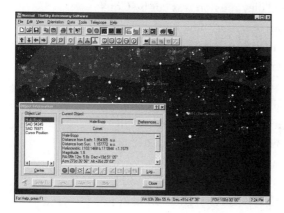

FIGURE 5.9
Comet Hale-Bopp as displayed by the
program The Sky

programs include orbits of well-known comets, and some include
more detailed databases of known comets. These programs, which
include Distant Suns (Figure 5.8) and The Sky (Figure 5.9), both
for Windows, will show you the positions of comets in the night
sky, although with few exceptions you will not be able to see the
comets with the naked eye, or even anything less than a research-
class telescope.

Fortunately, you can add the orbits of new comets to these programs. You will need the comet's orbital elements, which are
usually provided by the Minor Planets Center at the Harvard-
Smithsonian Center for Astrophysics. Check out its International
Comet Quarterly at **http://cfa-www.harvard.edu/cfa/ps/icq.html**

for up-to-date information about comets and links to other resources at its site and elsewhere. *Sky and Telescope*'s comets page at **http://www.skypub.com/sights/comets/comets.html** also has detailed information on current comets, including orbital information and viewing tips. Charles Morris, a longtime comet observer, provides his own comet observation information at **http://encke.jpl.nasa.gov/**; it is kept up to date as new comet discoveries occur.

Like asteroids, comets have not been studied in great detail by spacecraft despite their considerable numbers. In an international effort to send spacecraft to study Halley's comet in 1986, spacecraft from Japan, the Soviet Union, and the European Space Agency (but not the United States, whose Halley mission was cut by Congress during budget cuts several years earlier) flew by the comet, radioing back information about it and its long tail. Perhaps the most successful was the European mission Giotto, which flew within a few hundred miles of the comet's nucleus. Despite being banged up by fast-moving dust grains near the nucleus, it returned our best images to date of a comet nucleus (Figure 5.10). Halley's nucleus turned out to be a small, surprisingly dark, potato-shaped mass of ice and dust. The National Space Science Data Center (NSSDC) has the best online summary of Giotto's mission to Halley's comet at **http://nssdc.gsfc.nasa.gov/planetary/giotto.html**.

Both the United States and Europe are planning new missions to comets. NASA is working on Stardust, an innovative low-cost mission that will fly through the tail of comet Wild five years after its 1999 launch, collecting dust samples. The spacecraft will return those samples to Earth, parachuting them to a landing in the Utah desert in 2006. The Jet Propulsion Laboratory (JPL) has the official Web site for the mission at **http://stardust.jpl.nasa.gov/**, which includes mission plans and educational information. The European Space Agency (ESA) is planning its own mission, Rosetta, which will launch in 2003 on a mission to comet Wirtanen. The spacecraft will rendezvous with the comet and drop a lander onto the comet nucleus. The ESA Web site has more details on the mission at **http://sci.esa.int/rosetta/**.

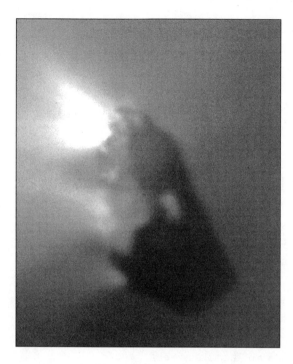

FIGURE 5.10
An image of the nucleus of Halley's comet
from the NSSDC Web site

Halley, Hyakutake, and Hale-Bopp

The most famous known comet is certainly Halley's comet. Its appearances every 74 to 76 years have been traced back in the historical record to at least 240 B.C.E. This large comet usually makes a bright appearance in the night sky when it passes though the inner Solar System, with the notable exception of its last passage in 1985–86. There are general and scientific resources about the comet at **http://seds.lpl.arizona.edu/nineplanets/nineplanets/halley.html** and **http://spaceart.com/solar/eng/halley.htm**.

While the last passage of Halley's comet through the inner Solar System was not a brilliant display as seen from Earth, we have been fortunate in the last couple years to have two bright comets appear in the night skies. In spring 1996, comet Hyakutake made a brief but brilliant appearance in the night sky as it passed near Earth. In 1997, comet Hale-Bopp, which was discovered the year before

Hyakutake in the outer Solar System, made an even more impressive appearance, outshining all but the brightest stars and planets for several weeks.

Comet Hyakutake was a sudden, and very pleasant, surprise. The comet was found in late January 1996 by Yuji Hyakutake, an amateur astronomer in Japan. Astronomers used a few days of observations of the comet to plot its orbit and determine the comet would pass close to the Earth, at least in astronomical terms: perhaps as close as 10 to 15 million miles. This meant that even if comet Hyakutake was a small, ordinary comet, it could be a brilliant sight in the night sky for a few days around its closest approach.

For observers in the Northern Hemisphere, who got the best view of the comet, it did not disappoint. Hyakutake become one of the brightest objects in the night sky, and its tail stretched behind the comets for dozens of degrees for those who could observe away from light-polluted skies. This magnificent sight lasted for only a few weeks, though, as the comet sped by the Earth and swung around the Sun to travel back out to the far reaches of the Solar System.

There are several online sources of information about comet Hyakutake. The Harvard-Smithsonian Center for Astrophysics has a page of basic information about the comet at **http://cfa-www. harvard.edu/cfa/ps/pressreleases/info1996B2.html**. *Sky and Telescope* magazine also has some information at **http://www.skypub.com/ comets/hyaku2.html**.

During Hyakutake's close approach, professional and amateur astronomers took thousands of images of the comet. Many of these images made their way to **http://www.jpl.nasa.gov/comet/ hyakutake/**, the largest online collection of images of Hyakutake. There are over a thousand images and animations of the comet archived here, organized chronologically. This site also has a list of links to other comet Web sites at **http://www.jpl.nasa.gov/comet/hyakutake/ other.html**.

Hyakutake, though, turned out to be just a warm-up for the main event, comet Hale-Bopp. Astronomers Alan Hale and Thomas Bopp both found the comet on the evening of July 22,

1995. The two amateurs were working independently of one another, Hale observing from the driveway of his New Mexico home and Bopp observing with a group of friends in the Arizona desert. It turned out that at the time of their discovery, the comet was still farther away from the Sun than Jupiter, a sign that the comet was very large and bright, and thus promised to put on an impressive show when it passed through the inner Solar System in early 1997.

Despite concerns that the comet was being overhyped and could not possibly meet the expectations of astronomers and the general public, the comet did not disappoint when it passed through the inner Solar System in early 1997. By April 1997, it reached a peak magnitude of nearly -1, making it one of the brightest objects in the night sky and easily visible from even the worst light-polluted skies. For those who had the benefit of darker skies, the comet was truly a spectacular sight.

Hale-Bopp encouraged even more people to take images of the comet and place them online. The largest collection of images was at the Jet Propulsion Laboratory, which accumulated over 5,100 images of the comet from amateur astronomers around the world by early May 1997, when the comet began to disappear from sight for many Northern Hemisphere observers. The site, at **http://www. jpl.nasa.gov/comet/**, includes images and animations of the comet, organized chronologically, similar to the archive of Hyakutake images mentioned above (Figure 5.11). NASA also established its own archive at **http://comet.hq.nasa.gov/**. This site was created in an effort to provide near real-time images of the comet as observers took images with their digital charge-coupled device (CCD) cameras and immediately uploaded them to the site, where they would be instantly accessible to visitors. Even though that's no longer happening, the site is still a good archive of images, albeit much smaller than the JPL collection.

Hale-Bopp became probably the first comet with its own Internet domain name and Web site, at **http://www.halebopp. com/**. Established with the assistance of Alan Hale and Thomas

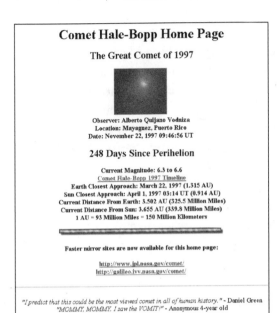

Comet Hale-Bopp Home Page

The Great Comet of 1997

Observer: Alberto Quijano Vodniza
Location: Mayaguez, Puerto Rico
Date: November 22, 1997 09:46:56 UT

248 Days Since Perihelion

Current Magnitude: 6.3 to 6.6
Comet Hale-Bopp 1997 Timeline
Earth Closest Approach: March 22, 1997 (1.315 AU)
Sun Closest Approach: April 1, 1997 03:14 UT (0.914 AU)
Current Distance From Earth: 3.502 AU (325.5 Million Miles)
Current Distance From Sun: 3.655 AU (339.8 Million Miles)
1 AU = 93 Million Miles = 150 Million Kilometers

Faster mirror sites are now available for this home page:

http://www.jpl.nasa.gov/comet/
http://galileo.ivv.nasa.gov/comet/

"I predict that this could be the most viewed comet in all of human history." - Daniel Green
"MOMMY, MOMMY, I saw the VOMIT!" - Anonymous 4-year old

FIGURE 5.11
The Hale-Bopp Web site at JPL provides easy access to thousands of Hale-Bopp images and animations, as well as updated news about the comet

Bopp, this site established itself as the comet's home page, with updated comet information, links to other resources, and a section of the site dedicated to stamping out some of the unfounded rumors and misinformation about the comet and any alleged companion a few observers reportedly saw.

Given the huge interest in the comet, and the increasing number of people accessing the Internet, hundreds of Hale-Bopp Web sites were created to show images, provide observing information, show research results, and more. The most comprehensive index to Hale-Bopp Web sites is at **http://www.seds.org/spaceviews/ halebopp/links/**, part of the SpaceViews space publication Web site. There are over 300 links to other Hale-Bopp Web sites here, cross-referenced in several categories such as images, information, observing tips, and Web sites in other languages.

In the last several chapters we have focused on the myriad of bodies in our Solar System, from our own planet to the Moon to other planets and smaller bodies. However, we have neglected until now the most important object in the Solar System, without which it would not exist: the Sun. The next chapter explores our knowledge of the body that is the heart of our Solar System.

FIGURE 5.12
The Hubble spots material ejected from Comet Hale-Bopp in these 1995 photos

6

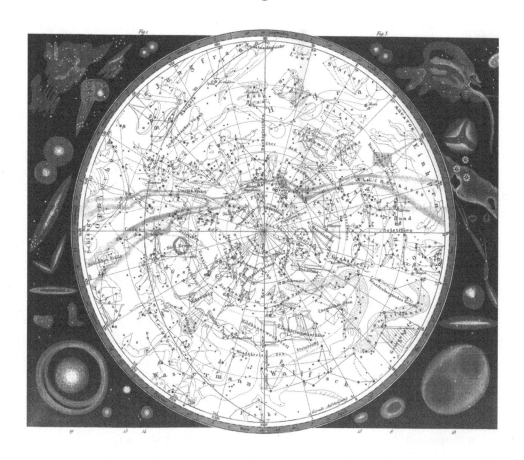

THE SUN

F OR MANY ASTRONOMERS, THE SUN is the enemy. It sets later in the evening than we desire, and rises earlier in the morning than we'd like. Its terrifically bright light, scattered by Earth's atmosphere to create the blue sky, makes it impossible to see other stars in the sky while the Sun is up. And the Sun, at first approach, seems to be a featureless, unchanging body.

Of course, that doesn't mean the Sun isn't interesting. We've come to learn that the Sun is not a bland, unchanging world. In fact, the Sun is very dynamic, with features on its disk, such as sunspots and solar flares—giant storms that change from day to day and can even have an impact on the Earth. In combination with the Earth's atmosphere and magnetic field, the Sun is responsible for the occasional formation of aurorae. The Sun is also responsible, through no work of its own but by a happy coincidence with the Moon, for fascinating eclipses as the Moon temporarily shades parts of Earth from the Sun.

This chapter will explore our Sun in detail, pointing out resources on the computer and online for more information and some of its fascinating features and effects.

The Basics of the Sun

O nly in the last few decades have we gained a better understanding on how the Sun works, thanks to observations from Earth and space, as well as improved knowledge of nuclear physics. For example, the source of its energy was a mystery until well into this century, when our growing knowledge of the Sun's composition and advances in our understanding of how atoms and subatomic particles interact, gave physicists the key information necessary to show that the fusion of hydrogen into helium created the tremendous energy necessary to power the Sun.

We now understand that the Sun is a giant sphere of hot gases, almost entirely hydrogen and helium, lacking any kind of conventional solid surface like a terrestrial planet's. The light we see comes from a region of the Sun called the *photosphere*. While one might think of the photosphere as the Sun's surface, it's actually a very tenuous region of gases, with a pressure as little as 0.01 percent that of the Earth's atmosphere at sea level! It is the effective surface, though, since the light coming from this region prevents us from looking any deeper into the Sun. The photosphere is usually thought of as the lowest level of the Sun's atmosphere.

Above the photosphere is a region called the *chromosphere*, a thinner, dimmer portion of the Sun's atmosphere. It has a pinkish appearance because the gas here is somewhat cooler than that in the photosphere (about 4,000 degrees Kelvin instead of 5,800), which means the light here is strongest at redder wavelengths. Usually you can't see the chromosphere with the naked eye except during an eclipse, when the Moon blocks the light from the brighter photosphere.

Above the chromosphere is an even more tenuous region of gas extending for millions of miles from the Sun, called the *corona*. Although one would expect the corona to be even cooler than the chromosphere, it is actually much hotter, with temperatures as high as two million degrees Kelvin. No one knows why it is so hot, although speculation centers around interaction of the gases in the corona with the Sun's powerful magnetic field. The corona is about as bright as the full Moon, but since it has to compete with the far brighter photosphere, it is not visible to the naked eye except during eclipses, when it appears to form an irregular halo around the Moon's darkened disk.

Since we cannot see past the photosphere, the structure of the Sun's interior is not as well understood. We do know that the fusion processes that generate the Sun's tremendous energy take place in its core, where pressures and temperatures are at a maximum. The energy created by the fusion reactions is then radiated away from the core and transferred to the photosphere through convection—just as hot gas rises, then falls when it cools.

There is a lot of basic information about the Sun available online. Naturally the two major Web sites about the solar system have devoted sections to it, with basic and many images taken from Earth and from space. Visit The Nine Planets (**http://seds.lpl.arizona.edu/nineplanets/nineplanets/sol.html**) or Views of the Solar System (**http://spaceart.com/solar/eng/sun.htm**) to get a basic introduction to the Sun.

Another excellent site dedicated to providing an educational introduction to the Sun is A Virtual Tour of the Sun at **http://www.astro.uva.nl/~michielb/od95/** (see Figure 6.1). This site goes through our knowledge of the Sun piece by piece, and uses a large number of images and videos to illustrate its phenomena. Based in the Netherlands, it is available in English and Dutch editions. The Sun: A Pictorial Introduction is an online slide show that provides some introductory information about the Sun, as well as a number of good images, at **http://www.hao.ucar.edu/public/slides/slides.html**.

If you're just interested in looking at some pictures of the Sun, The Planetary Society features an online gallery of solar images as part of the Web site, at **http://planetary.org/library/gsp-starssun.html** (see Figure 6.2). However, if you want some more in-depth information about how it works, Nick Strobel, an astronomy professor in California, has placed his lecture notes about the Sun from his astronomy class online at **http://www.bc.cc.ca.us/programs/sea/astronomy/starsun/strsuna.htm**.

If that's too serious, though, stop by **http://antwrp.gsfc.nasa.gov/htmltest/gifcity/interv.html** for a very funny "interview" with Sol, our Sun, and find out why dear old Sol complains that "sometimes I break out in those unsightly sunspots."

You don't have to be online, though, to find good introductory information about the Sun. Most planetarium programs provide at least some information, as well as useful information about the times of sunrise and sunset. Redshift 3 for Windows and Macintosh goes

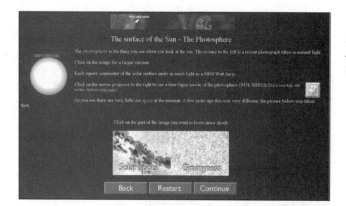

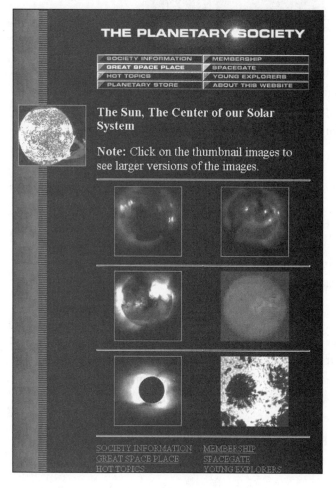

FIGURE 6.2

The Planetary Society's collection of Sun images

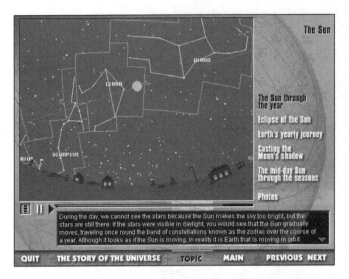

FIGURE 6.3
The path the Sun follows in the sky, from Maris's Discover Astronomy

even further with a guided tour, a several-minute narrated video which provides a good introduction. Discover Astronomy, by Maris Multimedia and available for Windows, has a section on the Sun that includes information about eclipses, the Sun's path in the sky, and some photos of the Sun from Earth and space (see Figure 6.3).

Solar Research

The Sun may be the biggest and brightest object in the sky, but it's difficult to learn much about it simply by using the naked eye. Just as with objects in the night sky, astronomers use telescopes to look at the Sun in greater detail to study features like sunspots on its surface and to look for solar flares. The National Solar Observatory, with branches in Arizona and New Mexico, is perhaps the best-known solar observatory. The distinctively shaped McMath Solar Observatory at Kitt Peak in Arizona is the world's largest solar observatory. You can learn more about it and its research online at **http://argo.tuc.noao.edu/**. The New Mexico branch of the National Solar Observatory, located in (where else?) Sunspot, New Mexico, also has a Web site at **http://www.sunspot.noao.edu/**. Here you can find information about the observatory and its research, as well as about the Sun in general and recent solar events.

Another good solar observatory is the Mees Solar Observatory, located atop a mountain named Haleakala on the Hawaiian island of Maui. Operated by the University of Hawaii, the observatory has been providing images of the Sun for many years. Its Web site, at **http://koa.ifa.hawaii.edu/**, is a good source of information about the observatory and the Sun. It offers many useful links to other solar observatories and research sites.

Although the Sun is easy to observe from Earth, spacecraft missions have provided more, higher-quality observations of solar phenomena and given us a better understanding of how the Sun works. One of the best known missions is Ulysses, a joint European-American mission to explore the Sun's poles, which cannot be seen well from Earth. Launched by the Space Shuttle in 1990, the spacecraft used the planet Jupiter to provide a gravity assist to allow it to fly back toward the Sun, flying first under the Sun's south pole and then above its north pole. The now-complete mission's Web site is at JPL, **http://ulysses.jpl.nasa.gov/ULSHOME/ulshome.html**. Although the spacecraft did not have a camera with

---

It is remarkable what sorts of instruments are available to the amateur astronomer today. Devices that only a few years ago fell firmly in the province of the professional astronomer are available today for amateur instruments. For example, specialized and sophisticated Hydrogen Alpha filters that attach to amateur telescopes are available, allowing the amateur to observe solar prominences. This is, for the amateur astronomer, a true "golden age."

There are some solar events, however, that don't require special equipment to observe — in fact, they don't occur on the Sun at all, but rather in our atmosphere. There is a rare event called the "Green Flash" that can occur when the Sun is setting — a product of the refraction of light through the Earth's atmosphere that causes a momentary flash of green on top of the Sun's orb as it sets. While I've seen photographs of the "Green Flash," I've never actually seen it with my naked eye — although I've looked often enough.
—R.L.

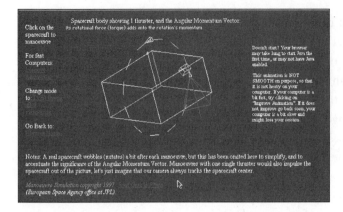

FIGURE 6.4
Control the orientation of the Ulysses spacecraft with this Java applet

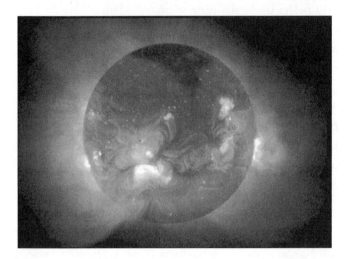

FIGURE 6.5
The Yohkoh spacecraft Web site features many X-ray images of the Sun taken by the Japanese satellite

which to take images of the Sun, there is other data about the Sun's magnetic fields, as well as some interesting Java applets that give you a chance to try controlling the Ulysses spacecraft (see Figure 6.4).

Another joint European-American mission is SOHO, the Solar and Heliospheric Observatory. This spacecraft is located in an orbit around the Sun near the Earth, returning images and other information about the Sun daily. Despite some recent problems with the spacecraft, it has become a very useful tool for spotting solar flares and other events on the surface of the Sun. The spacecraft has also witnessed other unusual events, such as a comet that came so close to the Sun it vaporized, if it didn't hit the Sun itself. The SOHO Web site, **http://sohowww.nascom.nasa.gov/**, provides information and interesting images of the Sun, and is updated regularly, especially when storms or other events are taking place on the Sun.

Japan's Yohkoh satellite studies the Sun at the wavelengths of mildly energetic, or soft, X-rays. That allows astronomers to get a different view of the Sun, with an emphasis on studying its interaction with its corona. The project has a Web site in the United States at **http://www.lmsal.com/SXT/homepage.html**, which includes more information and images from the spacecraft (see Figure 6.5). Also worth visiting is a special public outreach project that scientists and engineers in the United States have organized, at **http://www.lmsal.com/YPOP/**. They are using data from the Yohkoh spacecraft to explain to the general public, and to students, how the Sun works.

Another lesser-known spacecraft mission is Spartan 201, a satellite that has been temporarily deployed in Earth orbit several times. Brought up and returned to Earth by the Space Shuttle, most recently on the famous STS-95 mission that featured the second spaceflight of John Glenn, Spartan takes images and records other data about the Sun during its brief stay in orbit. Some of these images, as well as more information about the mission, are available online at **http://umbra.nascom.nasa.gov/spartan/**.

Sunspots and Flares

We take the Sun's light for granted: It always comes each day at the same strength. We don't see the amount of light fluctuate from day to day, or even year to year, so we come to believe that the Sun is a quiet, tranquil place, with only a few changes taking place over long periods of time. That couldn't be less true.

In fact, the Sun is a dynamic, changing body, whose appearance can change over the course of days. We don't ordinarily recognize these changes, since they are very small compared to the Sun's scale and don't noticeably alter the amount or quality of the light emanating from it, but with even a small telescope viewers can often see dark patches—sunspots—and on occasion giant storms—solar flares—erupting from the surface, sometimes sending a stream of charged particles towards the Earth.

Although there are some records of giant sunspot features the Chinese saw on the Sun 2,000 years ago, regular observations of sunspots did not begin until the early seventeenth century with the advent of the telescope. Early observers like Galileo and Christoph Scheiner (whose studies preceded Galileo's but aren't as well known) noticed that dark spots would occasionally appear on the solar disk and slowly move across it, sometimes changing size and shape. A good historical background on the study of sunspots is available online at **http://es.rice.edu/ES/humsoc/Galileo/Things/ sunspots.html**.

Twentieth-century astronomers have linked sunspots to the Sun's magnetic field. Sunspots appear where the field lines protrude through the photosphere. The field lines block rising hot gas, so the gas in the area of the photosphere around the field lines cools. This makes that region cooler and darker, and gives sunspots their characteristic dark appearance. By counting the number of sunspots over many years, astronomers have found that the number of sunspots follows an 11-year cycle, peaking in some years—for example, 1979 and 1990—but reaching their minimum during in-between years. No one is sure why sunspots follow that pattern, which has been observed for more than 100 years.

The Web has many sites with information about sunspots. The University of Oregon provides some images and explanation of sunspot phenomena at **http://zebu.uoregon.edu/~soper/Sun/ sunspots.html**. The Athena project for earth and space science education for kindergarten through 12th grade has an introduction to sunspots tailored for students, at **http://inspire.ospi.wednet. edu:8001/curric/space/sun/sunspot.html**. The Jet Propulsion Laboratory has an Educator's Guide for Sunspots online at http://learn.jpl.nasa.gov/sunspots.htm, which explains what sunspots are and what activities students can do to learn more about them. Similar educational activities are provided by Florida State University at http://www.scri.fsu.edu/~dennisl/CMS/sf/sunspots.html.

If your interest lies in the more advanced research aspects of sunspots, you will likely want to visit the Sunspot Index Data Center at http://www.oma.be/KSB-ORB/SIDC/. This site provides a record of the sunspot index, the number of sunspots seen on the Sun on a given day. You can use this to track solar activity on your own, and perhaps even look for trends of correlation with other astronomical or meteorological events.

Sunspots aren't the only evidence for a changing, active Sun. In some cases, the Sun's magnetic field will compress and heat up a small portion of the photosphere to extremely high temperatures, as high as 5 million degrees Kelvin. These gases will then burst free of the surface in an intense but brief event known as a solar flare. Solar flares follow the same 11-year pattern of activity as sunspots, with more flares during times of high sunspot activity.

Solar flares are of interest to people on Earth since streams of charged particles accompany these flares. These particles can reach Earth and interact with its magnetic field. They may also cause electrical damage to the circuits of a few satellites in Earth orbit. The failure of an AT&T communications satellite in January 1997 has been linked to a solar flare event that took place a few days before. NASA's Goddard Space Flight Center has more information about solar flares, including some good background, at http://hesperia.gsfc.nasa.gov/sftheory/.

The program Beyond Planet Earth, by Discovery Channel Interactive, has a feature called The Electric Sun. This narrated video explains the Sun's dynamic nature and how sunspots and solar flares form (see Figure 6.6). It also discusses how these are related to the aurorae seen on the Earth, as we'll explain below.

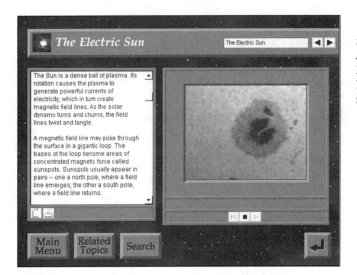

FIGURE 6.6
A screen shot from The Electric Sun, part of Beyond Planet Earth, describing the formation of sunspots

Aurorae

Those solar flares, which can disrupt satellites in orbit and in extreme cases disrupt electrical power systems on Earth, are also responsible for one of the most fascinating sights in the night sky at high latitudes: the aurorae. The Sun emits charged particles, called the solar wind, which are usually trapped by the Earth's magnetic field high above our atmosphere. At times—for example, during solar flare events—the wave of particles can overwhelm our magnetic field and get through to the upper atmosphere. There the particles collide with gas molecules, transferring their energy. This allows the gas particles to fluoresce, or glow, creating the aurora.

Aurorae are usually visible only at high latitudes, where the Earth's magnetic field drops down to the north and south magnetic poles. (The aurorae are known as the aurora borealis in the Northern Hemisphere and the aurora australis in the Southern Hemisphere.) During intense solar flare events, though, aurorae can be seen at latitudes much closer to the equator. Strong aurora events are linked to strong solar flares, so the best aurorae usually take place when the Sun is most active.

There are several good Web sites with more information about aurorae. The Athena educational project features some basic information about the aurorae, aimed at students, at **http://athena. wednet.edu/curric/space/solterr/sunaur.html**. The University of Alaska, one of the leading sites for auroral research, has some general information about the aurorae, in a site called The Straight Scoop: What Is the Aurora? at **http://dac3.pfrr.alaska.edu:80/ ~ddr/ASGP/STRSCOOP/AURORA/INDEX.HTM**. Michigan Technological University also has images and links to other sites at **http://www.geo.mtu.edu/weather/aurora/** (see Figure 6.7). There is more good information about the aurorae from Norway, in English, at **http://www.uit.no/npt/nordlyset/nordlyset.en.html** and from a magazine, also in English, at **http://www.uit.no/npt/nordly-set/waynorth/00-innhold.en.html**.

The aurorae, like the weather, change on a daily basis and are sometimes hard to predict. Online resources are often the best way to keep track of the Sun's condition and the possibilities for observing aurorae. The Solar Terrestrial Dispatch, **http://solar.uleth. ca/solar/**, provides information that's updated daily on the status of solar activity and the possibility for auroral or other activity on Earth. Today's Space Weather, at **http://www.sel.bldrdoc.gov/ today.html**, also provides daily information on the status of the Sun in the form of a forecast, with descriptions like this: "Solar activity

FIGURE 6.7
One of the many images of the aurorae available from the Michigan Technological University's Web site

95

is expected to be very low to low." This site also has a Space Environment Center, **http://www.sel.bldrdoc.gov/current_images. html**, which lets you view current and past images of the Sun in various formats to judge the level of solar activity for yourself. The Space Weather Bureau (**http://www.spaceweather.com**), a project of NASA's Marshall Space Flight Center, also provides updated information on solar activity as well as space science news and tutorials on the Sun and solar activity.

If you need more detailed information about aurorae, you can check in the Space Physics Textbook, an online textbook about the aurorae and other aspects of space physics, at **http://www.oulu.fi/ ~spaceweb/textbook/auroras.html**. This site has more information about the physical processes at work behind the aurorae.

Solar Eclipses

One of the most fascinating, and for many of our ancestors one of the most frightening sights in the sky is a total eclipse of the Sun. Eclipses occur when the Moon passes directly between the Sun and Earth. For a few minutes, in a limited area on Earth, the Moon blocks the full disk of the Sun, turning day to night or deep twilight. A total eclipse is the only time viewers can see the Sun's corona without the aid of special equipment; it's the bright, irregular halo surrounding the Moon's black disk. (It's worth noting here that totality, when the Moon *completely* blocks the Sun's disk, is the only safe time to look at the Sun with the unprotected eye. At all other times, including the moments leading up to and following totality, the Sun's intense brightness can cause severe, permanent eye damage unless you use solar filters or projection techniques to view it.)

Ancient civilizations often feared solar eclipses. They were usually seen as the work of an angry god, displeased with his peoples, or a monster devouring the Sun. People would go to great lengths, including sacrifices, to prove to their gods or other beings that they were worthy of the Sun once again. Naturally these efforts seemed to work every time.

Today, however, the situation is essentially the reverse of what it was in ancient times. People no longer fear eclipses, but will go out of their way to view them. Since total solar eclipses take place no more than a few times a year, and are visible over a limited region of the globe, this means people will spend thousands of dollars on eclipse expeditions to exotic, out-of-the-way locales, or take eclipse cruises into the middle of the ocean to be in the best location for viewing.

A number of Web sites offer good basic information about solar eclipses. *Sky and Telescope's* Sky Online Web site keeps an up-to-date list of upcoming eclipses at **http://www.skypub.com/eclipses/eclipses.shtml**. This site also includes tips for viewing eclipses and links to other Web sites. One of the most thorough resources online about solar eclipses is Fred Espenak's eclipse Web site at NASA's Goddard Space Flight Center, **http://sunearth.gsfc.nasa.gov/eclipse/eclipse.html**, with detailed information about upcoming solar eclipses and many links to other online resources about eclipses. The Space Data Analysis Center also has information about and great images of eclipses, at **http://umbra.nascom.nasa.gov/eclipse/** (see Figure 6.8).

FIGURE 6.8
An eclipse image from the Space Data Analysis Center shows the "diamond ring" effect, when the Moon blocks all but a small part of the Sun

The EarthView Eclipse Network, **http://www.earthview.com/**, also features information about eclipses. This site focuses on the knowledge of solar eclipse expert Bryan Brewer, author of the book *Eclipse*. Jeffrey Charles has been eclipse-chasing, or traveling around the world to observe eclipses, for many years, and he has documented his efforts on his Web site at **http://www. eclipsechaser.com/**. His site includes a diary of his. As additional proof that people will travel around the world to view solar eclipses are Web sites with pictures from a June 1983 solar eclipse observed in Indonesia at **http://www.users.globalnet.co.uk/~cyk/eclipse.html**, and information on an expedition to Mongolia in March 1997 to observe another solar eclipse at **http://www.crl.com/~zlater/**.

The planetarium program Redshift for Macintosh and Windows, also provides good information about eclipses, including several tutorials that allow the user to step through solar eclipses to see what is taking place at a specific time. It also explains the difference between a total solar eclipse, when the Moon's disk is large enough to completely cover the Sun's disk, versus an annular eclipse, when the Moon's disk is not quite large enough to cover the Sun, leaving a ring of light around the Moon. The difference between the two is caused by the slightly changing distance of Earth from the Sun and the Moon (see Figure 6.9).

A major solar eclipse in August 1999 was visible throughout much of western Europe. Many millions of Europeans who had not had the opportunity to observe an eclipse locally in decades saw this eclipse. There were already a couple of Web sites devoted to it, with maps of where the path of totality (the locations on the ground that will see the total eclipse) laid and recommendations for observing the eclipse. These sites are at **http://www.hermit. org/Eclipse1999/** and **http://www.users.dircon.co.uk/~seaview/ eclipseframe.htm**; the latter has a special emphasis on Great Britain. Those interested in past solar eclipses visible from Great Britain can check out **http://www.clocktower.demon.co.uk/eclipse99/**, which has information about a British book, *UK Solar Eclipses from Year 1*.

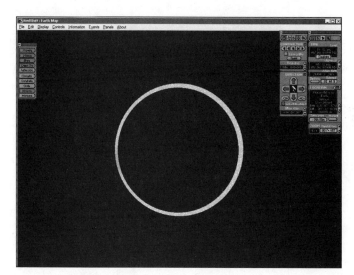

FIGURE 6.9
An animation in Redshift
shows an annual eclipse,
when a ring, or annulus, of
the Sun is visible around the
Moon

We've found in this chapter that the Sun is a far more lively,
dynamic body than we would have ever suspected. However, in the
greater scheme of things, it is a rather ordinary star, just one of
hundreds of billions in our galaxy alone. In the next chapter, we'll
look at some of the other stars in our galaxy, from tiny, dim red and
white dwarf stars to the massive supergiants.

7

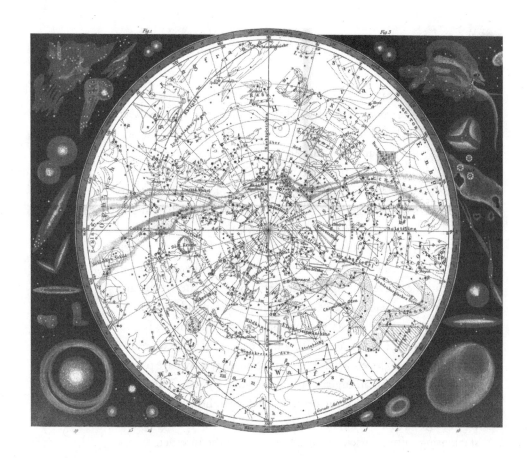

OTHER STARS

I F YOU HAVEN'T SEEN THE NIGHT sky on a clear, dark night far away from city lights, then you haven't truly seen it. The bright lights of cities keep urban dwellers from seeing more than a few hundred, or even a few dozen, of the brightest stars in the night sky. Go outside of town, though, and the sky looks radically different: Instead of a few bright stars and patches of darkness, the sky is studded with stars bright and dim. This is the sight that has captured people's attention and imagination for thousands of years.

While we've viewed stars in the night sky for millennia, only in modern times have we begun to understand what they really are. Not until the nineteenth century did people commonly accept the belief that stars were in fact versions of our own Sun, much farther away than any of the planets. Only in recent years have we found that not only are these stars like our own Sun, they may have planets around them, in an arrangement like the Solar System. This chapter will examine some of the properties of stars, how to find them in the night sky, and the possible existence of other worlds around stars, using software packages and resources on the Internet.

Stellar Basics

The stars we see in the night sky are suns like our own in the sense that they are giant balls of hydrogen and helium that create their energy through nuclear fusion. However, there's a wide range of stars that can vary greatly from our own run-of-the-mill Sun. Some are giant red or blue stars, many times larger and brighter than the Sun, while others are much smaller and much dimmer red and white dwarfs. Our Sun lies in the middle, one of many average stars.

Viewers can easily see part of this variation in the night sky. Some stars are much brighter than others, and they appear in different colors, including red, yellow, and white. Since ancient times, people have classified stars based on their brightness, or magnitude, a system that's been carried through to modern times with only a few changes. A star one magnitude brighter than another is about 2.5 times brighter, and smaller magnitude numbers represent brighter stars (this comes from the ancient practice of grouping the brightest stars together as stars of the first magnitude, slightly dimmer stars as stars of the second magnitude, and so on.). However, these magnitudes are dependent not only on the type of star, but on its distance from the Earth, so we needed a better way to classify stars.

Stars, per se, are not particularly interesting to observe through a telescope. Telescope size or level of magnification gains you nothing—no matter what instrument you use, a star still looks like a star. Yes, there are different colors of stars, and stars that vary in brightness, even stars that explode. Still, through the telescope, a star is a star is a star.

Now, if you start adding stars, things begin to get more interesting. Double stars, stars that are either in gravitational orbit about one another, or simply in a direct line-of-sight, can be interesting to observe. One beautiful example that comes to mind is Albireo, in the constellation of Cygnus. It's hard to imaging a more beautiful double star—one companion is golden, the other blue. Even after many years of observing, I still find this double star a fascinating and beautiful object to observe.

If you add even more stars, things get more interesting still. Whether it's a small open cluster of stars such as the M44, the "Beehive" cluster in the constellation Cancer, or M45, the famous "Pleiades" in the constellation of Taurus, clusters of stars are interesting to observe. Then, of course, there are the globular clusters, huge spherical congregations of stars such as the famous M13 cluster in Hercules.

While its true that some of our most powerful instruments can now observe some detail on some distant stars such as Betelgeuse in the constellation Orion and the variable star Mira in the constellation Cetus, I don't expect to possess such an instrument in my lifetime. So for me, a single star is best appreciated by my naked eye, and not through an optical instrument.
—R.L.

Starting early in this century, astronomers made a concentrated effort to classify the stars visible in the night sky. Led by astronomer Annie Jump Cannon at Harvard, they developed a classification scheme based on the spectrum, or distribution of light by wavelength, of each star. Other astronomers rearranged their original scheme of classes A through P in the 1920s as they were able to better understand the relationship between stellar spectra and their temperature. This led to the system used today, OBAFGKM, which has been immortalized with the mnemonic "Oh, Be A Fine Girl, Kiss Me" (or, depending on who you want to be kissed by, "Oh, Be A Fine Guy, Kiss Me"). This classification ranges from the blue-white O class stars, which have a surface temperature of over 35,000 degrees Celsius, to the dim M class stars, which are less than one-tenth as hot.

Also early this century, two astronomers working independently tried to link the spectral class of stars with their inherent brightness. Danish astronomer Ejnar Hertzprung and American astronomer Henry Norris Russell came up with a diagram known today as a Hertzprung-Russell, or H-R, diagram. It shows that when one plots *absolute magnitude* (a measure of a star's brightness that doesn't depend on the distance of the star from Earth) versus spectral class, a majority of stars fall on a curved line known as the main sequence. The diagram also shows three other classes of stars: a group of stars below the main sequence, known as *white dwarfs*; one above the red stars of the main sequence, known as *giants*; and one above the giants, the *supergiants*. A star on the main sequence will spend most of its life there, but may deviate from it late in life, as it uses up hydrogen fuel in its core.

To illustrate these differences in stars on the H-R diagram, you can use David Irizarry's simple shareware program for Windows called H-R Calc. This program displays an interactive H-R diagram; you can click on a particular location on the diagram and find out not only what the brightness and spectral class of that star would be, but also determine its size (see Figure 7.1). You can also adjust the temperature and brightness manually to see what effect it

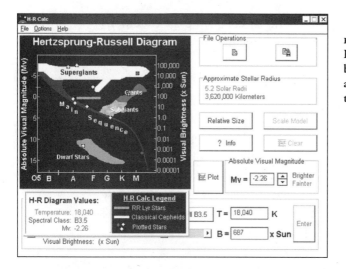

FIGURE 7.1
Finding the relationship between the size, brightness, and spectral class of a star with the program H-R Calc

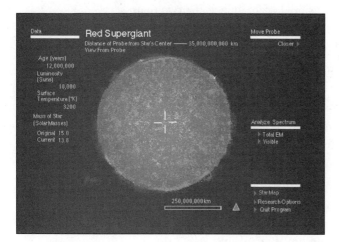

FIGURE 7.2
A close-up of the red giant star Betelgeuse from the program Star Probe

has on the star.

Another useful program, designed primarily for educational use, is Star Probe for the Macintosh, by Gemini Software. Star Probe allows the user to send spacecraft to explore specific stars, like the red giant Betelgeuse (see Figure 7.2). You can send the probe right into the heart of the star to uncover the physical processes at work, and see what conditions are like in the core of a giant star (a hint: it's *extremely* hot!). By comparing the conditions of various stars, you get a better understanding of how stars resemble one another and how they differ.

Double and Variable Stars

We tend to think that other stars are just brighter or dimmer versions of our own Sun: a solitary body, shining steadily. In fact, single unvarying stars may be in the minority in our galaxy. Many stars we see in the night sky are double and sometimes triple systems, with two or three stars orbiting a common center. Some stars are variable, with a brightness that changes over hours, days, weeks, or even years.

Double stars come in a number of varieties. Some consist of two stars of similar or near-similar size, such as the two main stars of the Alpha Centauri system. Others, like Sirius, have one dominant star and one much smaller companion that orbits the main star. When both members of a binary star system can be resolved in a telescope, this is called a *visual binary*. Not all visual binaries are actual binary star systems: In many cases two stars appear close to each other in the sky from Earth, but are actually many light-years apart. In some cases, only one member of the binary star system is visible, but the other can be detected indirectly through shifts in the star's spectrum. This is known as a *spectroscopic binary*. *Eclipsing binaries* are double stars where one star passes in front of the other as seen from Earth; this temporarily dims the light we observe, since part of the system's combined light is blocked.

Based on discoveries of binary stars already made, it's estimated that perhaps half the star systems in our galaxy may be binary. An online library of research information and references about double-star research is available at **http://www.chara.gsu.edu/DoubleStars/intro.html**. This site is sponsored by a research project at Georgia State University that is building a special interferometer system to make very high-resolution measurements of binary stars and other objects. The planetarium program Redshift, for Macintosh and

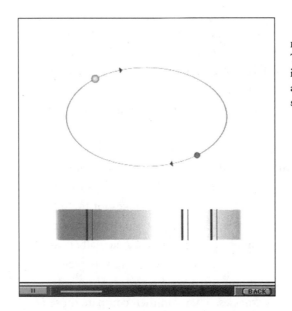

FIGURE 7.3
The binary star tutorial in Redshift explains how astronomers detect a spectroscopic binary

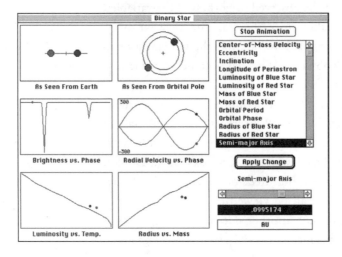

FIGURE 7.4
A user can adjust a number of different binary star parameters in the program Binary Star to see what effects they have on the system

Windows, has an excellent tutorial that explains the types of binary stars and how the stars interact with one another (see Figure 7.3). Another program, Binary Star for Macintosh, provides a basic introduction to binary stars and allows the user to model changes in the size and orbits of binary stars to see how those changes would appear to an observer on Earth (see Figure 7.4).

Variable stars, as the name suggests, are stars that change in brightness over regular periods of time. Variable stars are usually divided into two classes: RR Lyrae variables are low-mass stars whose brightness changes over periods under a day, while Cepheid variables are higher-mass stars whose brightness changes over days or weeks as they expand and contract.

Cepheid variables are of particular interest to astronomers, as they serve as a way of measuring the distance to distant galaxies. Cepheid variables follow a known period-luminosity relationship: If you know the period of the variable, the time it takes for the star to complete one cycle of brightness changes, you can determine its "intrinsic brightness"—the amount of light the star is actually generating. By observing changes in brightness of Cepheid variables in distant galaxies over time, astronomers can determine their periods, and thus their intrinsic brightnesses. They can then compare those intrinsic brightnesses with the observed brightness of the variables, which will be much smaller since the stars are very far away. Astronomers can then compute how far away the stars must be to have the brightness they observed. This has become a key method for finding the distance to faraway galaxies.

There are a variety of variable star resources online. The American Association of Variable Star Observers (AAVSO) has an introduction to the topic on its page titled What Are Variable Stars? at **http://www.aavso.org/variablestars.stm**. Canada's David Dunlop Observatory has a database of hundreds of known Cepheid variables in our galaxy, which can be viewed or searched online at **http://ddo.astro.utoronto.ca/**. Another database of Cepheids is online at **http://www.physics.mcmaster.ca/Cepheid/HomePage.html**.

Finding Stars and Constellations

Seeing stars in general is no problem: Even in the middle of the city, you can see dozens of stars, although the bright city lights will limit you to a handful of the brightest ones. However, if you're trying to find a specific star, picking it out from even just a few, assuming it's visible in the sky at the time, can intimidate a beginner.

One tool that has helped people locate stars for centuries is the constellations. These patterns of stars—88 in all, covering the northern and southern skies—can help you recognize particular stars. Most people are at least a little familiar with some constellations, such as Orion, or parts of them, like the Big Dipper. (This is not a constellation but an *asterism*, a well-known portion of a larger constellation, Ursa Major. The rest of the stars in Ursa Major are considerably dimmer than the stars in the Big Dipper.)

To use the constellations to find your way across the sky, though, you have to be able to recognize them. This is where the computer becomes very useful, with software and online resources providing a guide to the constellations and their component stars. Richard Dibon-Smith's The Constellations is a Web site that mixes the mythology and history of each constellation's name with information about the stars and other bodies within the constellation. There is a also a star map of each constellation (see Figure 7.5). You can check it out online at **http://www.dibonsmith.com/constel.htm.**

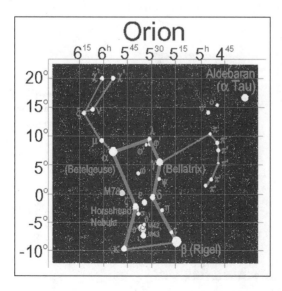

FIGURE 7.5
A star map of the constellation Orion, from The Constellations Web site

A site at the University of Wisconsin has information about constellations that includes a large list of stars, with information about each one, at **http://www.astro.wisc.edu/~dolan/constellations/**. Another site there has an even larger list of stars grouped by constellation at **http://www.ras.ucalgary.ca/~gibson/starnames/**. There's also an online pronunciation guide to constellation names at **http://www.eaglequest.com/~bondono/iconst.html**. *Sky and Telescope* magazine's Alan MacRobert provides some information on the origin of star names at **http://www.skypub.com/tips/basic/starnames. html**. An Annotated Guide to Constellation Resources on the Internet, at **http://www.ualberta.ca/~mbrundin/constellations/ index.htm**, provides references to other constellation Web sites and resources.

There are several sources online for creating and viewing star maps. One of the best is the set of interactive star maps produced by the National Geographic Society, at **http://www.nationalgeographic.com/features/97/stars/**. This site has a set of beautiful star maps you can use to find your way across the night sky (see Figure 7.6). The individual maps are also linked to Hubble Space Telescope images of some of the objects listed on them. You can generate customized star maps at a couple of sites online, including

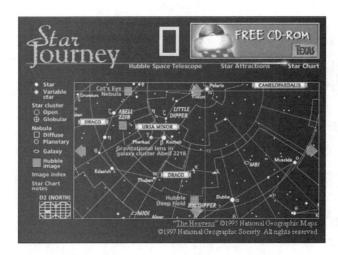

FIGURE 7.6
A star map of stars easily visible in the Northern Hemisphere, from National Geographic's online star map

http://www.polaris.net/services/starchart/ and http://www.mtwilson.edu/Services/StarMap/. In each case, you can specify the part of the night sky you want to look at and the level of detail the maps should have. You will need to have either a PostScript-compatible printer to print out the maps or a program such as Ghostview to view the maps on your computer screen.

If you're just interested in the positions of stars, you can use several catalogs online. The astronomy department at the University of Toronto has compiled a list of 314 of the brightest stars, down to about magnitude 3.5, at http://www.astro.utoronto.ca/%7Egarrison/oh.html. These stars are bright enough that you have a good chance of seeing them even in light-polluted city skies. If you need a much larger database of stars, NASA's Goddard Space Light Center has Sky2000, a master star database of many thousands of star positions. This catalog is used to create star catalogs for spacecraft missions. You can get more information on the stellar database and download the entire 140MB database, or compressed or smaller subsets of it (the database in ZIP format is 23.1MB) at http://fdd.gsfc.nasa.gov/attitude/skymap.html.

Almost any planetarium program for your computer will serve as a good guide to the night sky. Some programs, like Galaxy Guide for Windows, provide information on each constellation, including mythology related to the constellation's name and information on the constellation's stars and how to find it. There are even some nice illustrations of the constellations themselves (see Figure 7.7). Other programs, like The Sky for Windows, include detailed information on thousands of stars, including their location and brightness, spectral type, distance from Earth, and more (see Figure 7.8). You can also find this information in other programs, like Voyager II and Starry Night for Macintosh and Windows.

All of these online resources and programs treat the stars in a two-dimensional manner. All the stars are considered to exist on the celestial sphere for these purposes. This holdover from the geocentric model of a Earth-centered universe still works well for determining the relative positions of stars in the night sky, but it

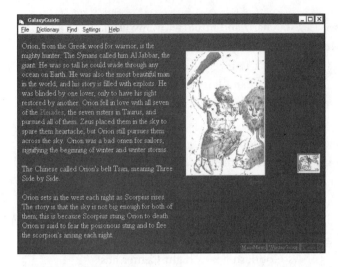

FIGURE 7.7
An illustration of the constellation Orion from the program Galaxy Guide for Windows

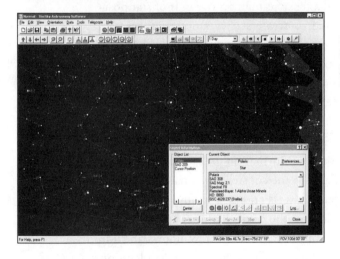

FIGURE 7.8
Some of the information the program The Sky provides for Polaris, the Pole Star

provides no information on their distance from Earth or the actual distance between them. The stars are in fact spread through three dimensions. To get a better feel for this, you can view some VRML (Virtual Reality Markup Language) models of the stars near the Sun at The 3-D Sky, **http://www.honeylocust.com/Stars/**. This site has 3-D models of the distribution of hundreds of bright stars near our Sun, which you can view in a Web browser or another program that supports VRML worlds (see Figure 7.9). After exploring some

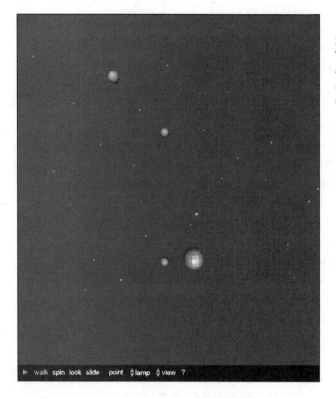

walk spin look slide   point   ⬦lamp ⬦view ?

FIGURE 7.9
This VRML model shows the
Sun (gray sphere) and some of its
nearest stellar neighbors (white
spheres)

of these models you'll get a better appreciation of how stars are distributed in our portion of the Galaxy.

Other Suns, Other Planets

We have long since recognized that the stars we see in the night sky are suns like our own Sun, and many astronomers have thought that at least some of them may have planets. The problem has been trying to find these planets; because they are much smaller than the stars they orbit, and are visible only by the light they reflect from their parent sun, in the past they were essentially impossible to observe from Earth. However, thanks to new observing techniques and better, more powerful instruments, we are beginning to discover new worlds around nearby stars.

The discoveries started in 1991, when Penn State University astronomer Alexander Wolzczan announced that three small planets were orbiting the pulsar PSR1257+12. Wolzczan used the giant radio telescope at Arecibo, Puerto Rico, to monitor changes in the timing of the radio pulses coming from the star to infer the planets' existence. Two appeared to have about the same mass as Earth, while the third had a mass similar to the Moon's. A Web page at Penn State's astronomy Web site, **http://www.astro.psu.edu/users/ pspm/arecibo/planets/planets.html**, has more information on the discovery and the techniques used to find the planets.

While the discovery of the pulsar planets was interesting, it was also a bit disappointing. It wasn't clear that the planets found around the pulsar had existed for the whole life of the star; they may have formed during or after the cataclysm that created the pulsar, or they may be remnants of much larger planets. Astronomers continued to search for planets around nearby Sun-like stars, and in the last two years they have uncovered a small bounty of new worlds.

The most prolific discoverers have been a team of astronomers led by Geoffrey Marcy at San Francisco State University and Paul Butler at the University of California at Berkeley. They have used indirect methods to discover several planets around nearby stars. Using a spectrograph attached to a telescope, they observe specific spectral lines that occur at fixed, known wavelengths. These lines can shift wavelength, though, if something is "tugging" on the star: The gravitational pull of a body like a planet can cause the star to wobble, creating a Doppler shift as the star wobbles toward and away from us, just as a car horn changes pitch depending on whether the car is speeding closer or going away. By looking for periodic variations in spectral lines, astronomers can determine how massive, and how far away from the star, the body must be to cause the observed Doppler shift.

Using this technique, Marcy, Butler, and their students have discovered or confirmed more than a half-dozen planets around nearby stars. All of these planets are large, in some cases having several times the mass of Jupiter, the largest planet in the Solar System. Many of these planets orbit much closer to their suns than Jupiter or the other massive gas giants orbit the Sun. This makes sense from an observational standpoint, since these solar systems create larger Doppler shifts and are thus easier to discover, but it gives us new insight into how solar systems form, and shows that not all solar systems resemble our own. Marcy and Butler's group has created an extensive Web site at **http://www.physics.sfsu.edu/~gmarcy/planetsearch/planetsearch.html** with more information about each planet they have discovered, including summaries of their data and reports on scientific papers they have published. Another major extrasolar planet research group, led by Michel Mayor of the Geneva Observatory in Switzerland, also has a Web site at **http://obswww.unige.ch/~urdy/planet/planet.html** to report their work.

There are other efforts to search for planets around other stars. One good summary of research efforts and their results is Exoplanets: Planets around Other Stars at **http://www.ph.adfa.edu.au/ e-mamajek/exo.html**. This site provides updated information on discoveries of new extrasolar planets and the techniques used to find them. The Extrasolar Planets Encyclopaedia, at **http://www. obspm.fr:80/departement/darc/planets/encycl.html**, also provides thorough information, with more of an emphasis on distributing research results throughout the scientific community. Other Worlds, Distant Suns at **http://otherworlds.home.ml.org/** not only has updated information about extrasolar planets, but has some 3-D representations of some of the new solar systems using VRML to give you a perspective on these new worlds (see Figure 7.10).

Although these discoveries have used Earth-based telescopes, there are plans for future spacecraft missions to provide more and better data on new solar systems. The proposed Kepler mission would use a 1-meter (39-inch) telescope in orbit around the Sun to look for tiny changes in a star's brightness caused when a planet passes in front of it. The instruments on Kepler would have to be very precise; the planet would cause the star to dim by only a factor of 1/100,000! Although Kepler wasn't chosen in a recent round of mission selections, the mission's designers hope NASA will select it

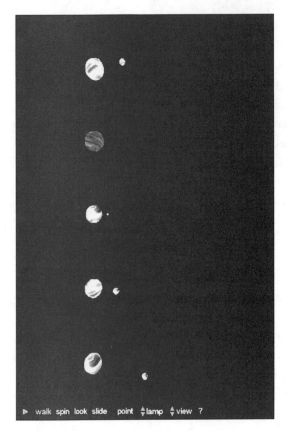

FIGURE 7.10
A close-up of some of the new solar systems using a VRML model available from the Other Worlds, Distant Suns Web site

in the future. You can read more about Kepler online at **http://www.kepler.arc.nasa.gov/**. The Jet Propulsion Laboratory has online the Exploration of Neighboring Planetary Systems, a road map for future plans to study nearby solar systems using first ground-based telescopes, and then a series of more capable telescopes in space. It's online at **http://techinfo.jpl.nasa.gov/WWW/ExNPS/HomePage.html**.

To date, all of our discoveries of extrasolar planets have been indirect: We have not seen the planets themselves, only the small changes they make in their parent star, through either Doppler shifts of spectral lines or changes in pulsar timing. Thus, we don't know what these worlds look like. That doesn't mean we can't speculate, though, and one artist, John Whatmough, has created some

FIGURE 7.11
Artist John Whatmough's rendition of the planet orbiting the star 51 Pegasi, just a few million miles from the star

impressive renditions. His Extrasolar Visions Web site at **http://www. empire.net/~whatmoug/Extrasolar/extrasolar_visions.html** includes some images of what he thinks some worlds, such as the planet closely orbiting the planet 51 Pegasi (see Figure 7.11), may look like. He's also included information about these worlds and how they were discovered. After seeing some of these images, you'll wish you could go and visit them for real.

While we have touched on how stars live and evolve, perhaps the most interesting aspect of the lives of stars is how they die. The more massive the star, the more impressive its death. Truly large stars can even explode as they run out of nuclear fuel. The next chapter will look at how stars end their lives, from the formation of red giants to supernovae.

8

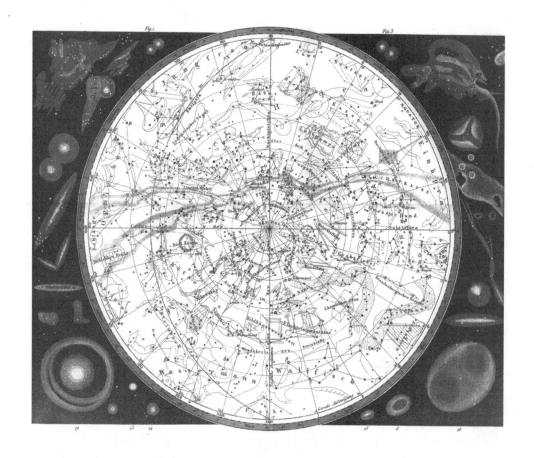

SUPERGIANTS AND
SUPERNOVÆ: THE DEATHS
OF STARS

FOR THOUSANDS OF YEARS, humans considered the night sky as an eternal, never-changing sight. The same stars appeared in the same positions from one year to the next, with the exception of a handful of wandering bodies—planets—that moved about the night sky in regular periods. The other stars, though, stayed in the same places and were always there, and would continue to be there forever.

Once in a long time, though—far less often than comets made their unexpected appearances—another new object would appear in the night sky, flare to great brightness, and for a few weeks be one of the brightest stars in the night sky. Then it would fade away and disappear, never to be seen again, leaving humans puzzling over its appearance. The Chinese called these apparitions *guest stars;* the most notable of these blazed in the night sky in the year 1054. Today we recognize that these guest stars seen in 1054, 1572, and 1604, among other years, are supernovae: sudden, tremendous explosions of distant stars that can briefly outshine even a whole galaxy.

We now know that supernovae represent one of the final stages in the life of a giant, massive star. Our observations of supernovae, and the deaths of other stars, help us understand how stars live, and how we can expect other stars, like our Sun, to act when their stellar lifetimes come to an end. In this chapter, we take a look at what happens to stars in old age as they exhaust their supply of hydrogen fuel. We especially emphasize how very large stars die, a process that includes one of the most violent and critical events in the universe: a supernova.

Stellar Evolution

Stars, like people, have particular lifespans: They are born out of clouds of gas and dust, live out most of their lives in a mature stage, and then die. The lives of stars are closely tied to their source of energy, nuclear fusion. A star's livelihood depends on its ability to generate energy by fusing atoms of hydrogen together to form helium, plus tremendous amounts of energy. Without this source of power the star would be unable to support its own weight under its overwhelming gravity, and would quickly collapse.

While all stars require nuclear fusion to stay alive, they generate energy at vastly different rates. One would expect more-massive stars, with their larger reserves of hydrogen, to last longer than smaller stars. However, the opposite is true: The more powerful gravity in larger stars increases the pressure in their core, where fusion takes place, and this increases the rate of fusion. A massive blue or white supergiant may last just a few million years, while a miserly red dwarf may continue fusion for tens of billions of years.

Supernovae that are visible to the naked eye are quite rare, but such exploding stars are can be seen fairly regularly with optical aid. While such stars are of great scientific interest, visual observations of such cataclysmic events leave much to be desired. Much more interesting visually are the remnants of these explosions, vast and intricate bubbles of gas that are ejected from the central event at great speed.—R.L.

An excellent interactive tutorial of this is available on the CD-ROM The Universe: From Quarks to the Cosmos, part of the Scientific American Library and available for Windows and Macs. Two sliders let you control the mass of the star and the point it is at in its life (see Figure 8.1). A brief examination of this will show that the less massive the star, the longer it will live. This program also gives you an idea of how stars of different masses die, which we will discuss in more detail.

For more online information about the lives of stars, check out a section on stellar evolution at **http://zebu.uoregon.edu/textbook/se.html**. This site, part of the Electronic Universe astronomy hypertextbook project at the University of Oregon, has more information and illustrations of the various stages in the life of a star.

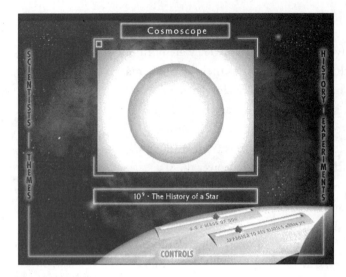

FIGURE 8.1
Explore the life stages of stars of different masses in The Universe, from the Scientific American Library

The Deaths of Normal Stars

The mass of a star determines, not only how long it lives, but how its life comes to an end. The death throes of most stars, including the Sun, are not spectacular, but are rather interesting.

At a point late in a star's life (about 5 billion years from now for our Sun), it runs out of hydrogen in its core to fuse, having converted all of it into helium over billions of years. Without this source of energy, the layers of the star around the core begin to contract and heat up. Soon a new shell of hydrogen fusion starts around the old, dormant core. Since this area of fusion is closer to the star's surface than the original core is, energy can travel more efficiently to the surface, heating it up and causing expansion. The star has become a *red giant*.

The red giant is much brighter than the original star (2,000 times brighter for stars like our Sun) and much larger: When our Sun becomes a red giant it will swell, by some estimates, out to the orbit of the Earth, or beyond. Red giants also begin to shed gases from their outer layers, as the force of gravity is weaker farther from the star's core, and the gases' energy has increased.

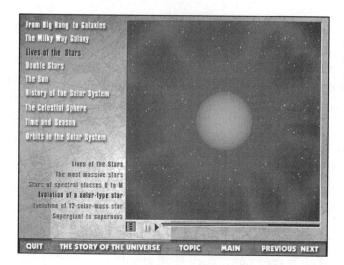

From Big Bang to Galaxies
The Milky Way Galaxy
Lives of the Stars
Double Stars
The Sun
History of the Solar System
The Celestial Sphere
Time and Season
Orbits in the Solar System

Lives of the Stars
The most massive stars
Stars of spectral classes B to M
Evolution of a solar-type star
Evolution of 12-solar-mass star
Supergiant to supernova

QUIT   THE STORY OF THE UNIVERSE   TOPIC   MAIN   PREVIOUS NEXT

FIGURE 8.2
A simulation of the red giant phase in the life of the Sun, from Discover Astronomy

A good computer tutorial that explains the red giant stage is available in the program Discover Astronomy for Windows. The animation shows how a Sun-like star swells into a red giant late in its life, expanding to Earth's orbit and shedding mass along the way (see Figure 8.2). Information about the lives of red giants is available online, in many cases as notes from college astronomy classes, such as http://plabpc.csustan.edu/astro/stars/giant.htm.

Eventually temperatures and pressures in the core grow to the point where helium begins to fuse, creating carbon and oxygen through a complicated process. This stabilizes the star and actually causes the star to grow dimmer, but hotter. However, the stability is temporary; for a star like our Sun, helium fusion will last only a couple billion years before the supply of helium is exhausted.

As was the case earlier, when hydrogen fusion began in a shell around the core, helium fusion begins again. This causes the star to become an even larger and more massive red giant, 10,000 times brighter than the Sun, expanding to a diameter as great as the orbit of Mars. The core grows hotter and the pressure there increases, but it is not enough to start a new round of fusion in the core. Fusion continues in the helium shell until a series of flashes, or sudden bursts of energy, blow away the star's outer layers.

The hot outer layers of the star form a ring of slowly cooling gases called planetary nebulae. The name is a misnomer: the first planetary nebula was discovered by English astronomer William Herschel, who thought that the ring around a bright core looked like a distant planet. A star can expel up to 60 percent of its mass into a planetary nebula.

The core of the star, composed of carbon and oxygen that isn't hot and dense enough to fuse, slowly cools. The star is now called a white dwarf. A white dwarf created from a star like the Sun will have at least 40 percent the mass of the original star, but compressed into a sphere no larger than Earth. A single cubic centimeter of white dwarf material would weigh a thousand tons.

The star spends the rest of its years slowly cooling off and dimming, no longer able to generate energy and supported from further collapse only by the pressure of electrons within it.

The Death of Giant Stars: Supernovae

Low-mass stars, like the Sun, end their red giant phases after fusing helium into carbon and oxygen because they lack the pressures and temperatures needed to start another round of fusion. However, a heavier star with several times the Sun's mass is capable of starting another round of fusion. Such stars continue to expand, becoming supergiants, like the star Betelgeuse.

The fusion process continues by fusing carbon, neon, oxygen, and finally silicon. The silicon fusion process creates as its main product atoms of iron. Iron, though, is as far as nuclear fusion can go; fusing atoms of iron takes away energy instead of producing it. The core of iron continues to grow until it reaches a critical mass, about 1.4 times the mass of the Sun. The stage is set for a truly cataclysmic event.

When the core reaches that critical mass, no force can continue to support it. In a fraction of a second, the core collapses until it reaches the same density as protons and neutrons packed together in an atomic nucleus, then rebounds. This creates shock waves that rip apart the star in a titanic explosion. The star has become a *supernova*.

Supernovae are among the most violent events in the universe. They can, for a short time, generate as much energy as all the other stars in a galaxy combined. Astronomers on Earth can easily observe supernova explosions in distant galaxies, as the supernova's brightness rivals that of its galaxy. A supernova in our galaxy can become a brilliant sight in the night, or even daytime, sky, as was the case with the 1054 explosion witnessed in China and Tycho's Star, the 1572 supernova the astronomer Tycho Brahe studied.

Many sources of information about supernovae are available online. England's Royal Greenwich Observatory has an information leaflet online with general introductory information about supernovae. It's available at **http://www.ast.cam.ac.uk/pubinfo/ leaflets/supernovae/supernovae.html**. NASA's Goddard Space Flight Center also has basic information, including an animation illustrating what happens during a supernova explosion, at **http:// legacy.gsfc.nasa.gov/docs/snr.html** (see Figure 8.3). A time line of studies of supernovae, as well as related objects, is online at **http://www.gsu.edu/other/timeline/dwarfs.html**.

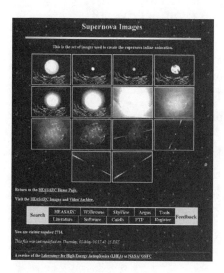

FIGURE 8.3
Scenes from an animation of a supernova explosion, from NASA's Goddard Space Flight Center

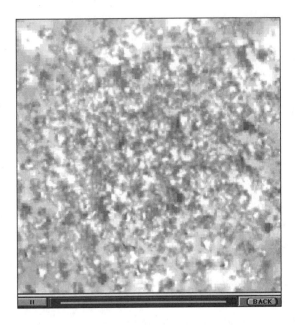

FIGURE 8.4
An animation of a supernova explosion from the program Redshift 2

Offline, several programs have information about supernovae. The CD-ROM Redshift 2 for Macintosh and Windows includes a tutorial on the lives of stars that focuses on the supergiant stage and supernovae (see Figure 8.4). Another program, Galaxy Guide, has some basic information on several supernova-related topics, including the 1987 supernova in the nearby Large Magellanic Cloud, which observers in the Southern Hemisphere could see with the naked eye, and the 1054 explosion that created the Crab nebula (see Figure 8.5). The program Distant Suns for Windows and Macintosh has some very basic information about Tycho's Star.

Because supernova explosions are unpredictable, detecting them requires astronomers to keep an eye on the entire sky. This is one area, like the discovery of comets, where amateurs can play a major role. While many supernova discoveries are made at major observatories, dedicated amateurs have also bagged their share of new supernovae. One such amateur is Californian Wayne P. Johnson, nicknamed Mr. Galaxy, who has discovered five supernovae. His Website, at **http://www.chapman.edu/oca/benet/**

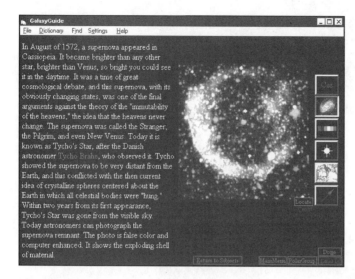

In August of 1572, a supernova appeared in Cassiopeia. It became brighter than any other star, brighter than Venus, so bright you could see it in the daytime. It was a time of great cosmological debate, and this supernova, with its obviously changing states, was one of the final arguments against the theory of the "immutability of the heavens," the idea that the heavens never change. The supernova was called the Stranger, the Pilgrim, and even New Venus. Today it is known as Tycho's Star, after the Danish astronomer Tycho Brahe, who observed it. Tycho showed the supernova to be very distant from the Earth, and this conflicted with the then current idea of crystalline spheres centered about the Earth in which all celestial bodies were "hung." Within two years from its first appearance, Tycho's Star was gone from the visible sky. Today astronomers can photograph the supernova remnant. The photo is false color and computer enhanced. It shows the exploding shell of material.

FIGURE 8.5

Information on the supernova of 1572 (Tycho's Star) is available from the program Galaxy Guide

mrgalaxy.htm, includes basic supernova information and observing tips. The International Supernovae Network, at **http://www.supernovae.org/isn.htm**, includes introductory information about supernovae, plus information on what to do if you discover one, including places to contact and what information to record. Observers' reports on supernovae in distant galaxies is available from the site Supernova in NGC and IC Galaxies at **http://www.ggw.org/freenet/a/asras/supernova.html**.

Although amateurs discover many of these distant supernovae, professionals still do the follow-up research. A good online resource on supernova research is the University of Texas supernova research group at **http://tycho.as.utexas.edu/**. This site has information on its research projects as well as a separate page of links to other supernova research projects and resources on the Web, at **http://tycho.as.utexas.edu/SN/links.html**. Another comprehensive list of links to research sites and supernova researchers is **http://cssa.stanford.edu/~marcos/sne.html**. While supernova research is usually related to observing supernovae and interpreting those observations, there is some theoretical research and computer modeling in progress on the topic. Information on one such effort

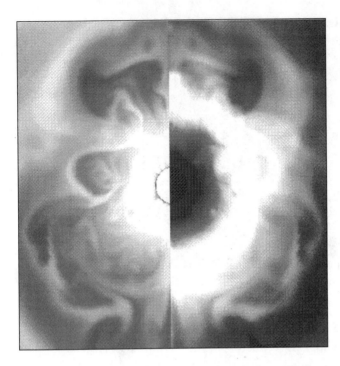

FIGURE 8.6
A scene from a computer simulation of a supernova explosion

is at **http://wonka.physics.ncsu.edu/www/Astro/Research/CSM/ supern.html**, which details efforts using supercomputers to model the physics of supernova explosions (see Figure 8.6).

There are extensive catalogs of supernova events available online. An Italian catalog, at **http://athena.pd.astro.it/~supern/**, has an extensive list of observed supernovae organized by date of discovery and location in the sky. The Harvard-Smithsonian Center for Astrophysics updates a list of recent supernovae at **http://cfa-www.harvard.edu/cfa/ps/lists/RecentSupernovae.html**. Another list of recent supernovae, as well as novae (sudden, periodic, nondestructive outbursts of stars not related to stellar evolution), is at **http://www.demon.co.uk/astronomer/variables.html**. A list of historical supernovae witnessed by humans over the last 2,000 years, most by the naked eye, is at **http://a188-l004.rit.edu/richmond/ answers/historical.html**.

While we lack the same dramatic images of supernovae that we have for objects within our own solar system, there are still online resources of good images. A site in Japan, **http://www. info.waseda.ac.jp/muraoka/members/seiichi/pictures/nova.html**, hosts images of recent supernovae and novae (see Figure 8.7). Some of the best images are of the aftermath of the 1987 supernova taken by the Hubble Space Telescope, **http://oposite.stsci.edu/pubinfo/ jpeg/SN1987A_Rings.jpg** (see Figure 8.8). These images show an unusual set of rings in an hourglass shape around the supernova. These rings are formed by gas ejected long ago from the star, illuminated by radiation. A ring of slowly expanding debris is also visible. When the expanding ring comes into contact with the hourglass sometime in the next few years, the shell will brighten considerably. Another good image is of the massive star Eta Carinae, **http://oposite.stsci.edu/pubinfo/jpeg/EtaCarC.jpg**, which seems to have suffered a supernova-like explosion in the nineteenth century yet managed to survive the event (see Figure 8.9). Lobes of gas ejected by the explosion make this star one of the most interesting objects the Hubble Space Telescope has ever imaged.

Considering the impressive power of supernovae, one question that may naturally come to mind is whether a supernova explosion could be dangerous to Earth. Such an explosion releases a tremendous burst of energy, including energetic, dangerous gamma rays.

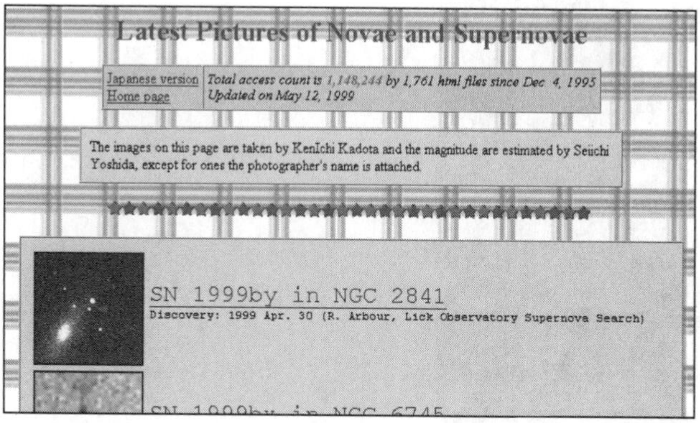

Latest Pictures of Novae and Supernovae

Japanese version Home page	Total access count is 1,148,244 by 1,761 html files since Dec. 4, 1995 Updated on May 12, 1999

The images on this page are taken by KenIchi Kadota and the magnitude are estimated by Seiichi Yoshida, except for ones the photographer's name is attached.

★★★★★★★★★★★★★★★★★★★★★★★★★★★★★★★★★

SN 1999by in NGC 2841
Discovery: 1999 Apr. 30 (R. Arbour, Lick Observatory Supernova Search)

SN 1999by in NGC 6745

FIGURE 8.7
Amateur astronomers' images of novae and supernovae are available online

FIGURE 8.8
A Hubble image of the 1987 supernova, including its unusual rings and hourglass shape

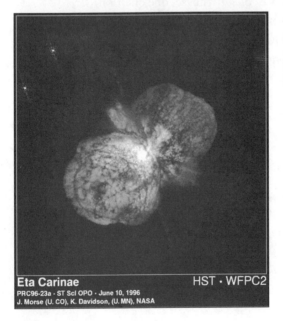

FIGURE 8.9
A Hubble image of Eta Carinae in the aftermath of a supernova-like explosion

This has been postulated as one possible cause for some mass extinctions in the Earth's history; perhaps the planet was irradiated by a nearby explosion. Astronomer Michael Richmond addresses these concerns at the site Will a Nearby Supernova Endanger Life on Earth? at **http://a188-lo04.rit.edu/richmond/answers/snrisks.txt**. It's something to think about, but not to lose sleep over.

While supernovae may pose a slight danger to life on Earth, they were also essential to creating it. During a supernova explosion, one final burst of nuclear fusion takes place in the star's exploding shells, creating dozens of different elements, including the ones beyond iron that do not release energy, and dispersing them into space. These elements, almost an afterthought of the supernova process, become the building blocks for planets and for people. We are, as they say, made of the stuff of stars.

Supernova explosions also play another critical role. The blast waves spread outward and can pass through clouds of interstellar gas and dust, creating clumps of matter large enough to attract other matter. These clumps begin to collapse and grow into protostars. The death of a star in a supernova can trigger the birth of many more stars.

While a supernova marks the star's death, its story is not yet over. A remnant of the star is left behind from the explosion. Unable to use nuclear fusion to power itself, it will begin to collapse. Depending on the mass of the body, this remnant will take on one of several forms that include some of the most exotic objects in the universe: neutron stars, pulsars, and the enigmatic black holes. The next chapter looks at these unusual, interesting objects left behind by supernovae.

9

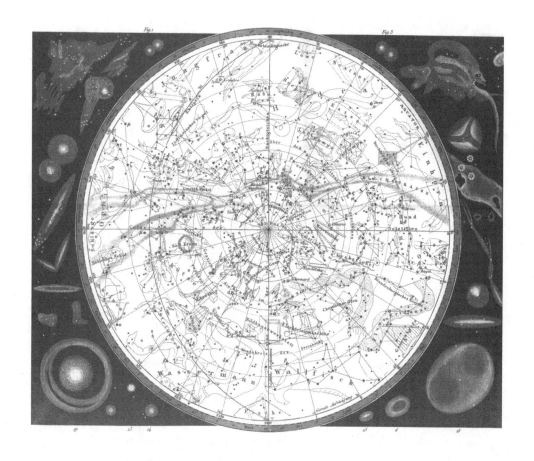

NEUTRON STARS, PULSARS, AND BLACK HOLES

WHEN VERY MASSIVE STARS—with four times the mass of our Sun or greater—are no longer able to generate energy through nuclear fusion, they explode in a titanic explosion called a supernova, as discussed in the last chapter. You might think such an explosion would mark the very end of that star's life, and in many respects it does, since the explosion blows away most of the star's mass. However, the explosion marks the beginning of one final life stage. The dense remnant of the old star's core, unable to support itself through nuclear fusion, will collapse further under its own gravity.

The collapse of these stellar remnants creates some of the most exotic objects in the universe, including neutron stars, densely packed cores of neutrons; pulsars, neutron stars with rotating magnetic fields; and black holes, objects whose gravity is so intense that even light cannot escape their pull. Black holes alone have such amazing properties that the term has entered common usage, though most people don't truly understand the astronomical phenomenon from which the name comes. After reading this chapter, though, you'll better appreciate black holes, neutron stars, and pulsars, as well as become familiar with resources about these objects online and in software.

Neutron Stars

When stars about four times as massive as our Sun explode, they leave behind a small remnant from their cores, usually with no more than about twice the Sun's mass. The stellar remnant begins to collapse under its own gravity. In less massive stars, like our Sun, this contraction stops when the electrons get close enough to repel one another, balancing gravity. This creates a white dwarf, which the Sun will become in about 5 billion years.

In more massive stars, however, gravity is so strong that it overcomes even the powerful force of electron repulsion. The density in this stellar core becomes so high that electrons combine with protons, forming electrically neutral particles called neutrons. The neutrons are pressed together so tightly that the pressure they exert on each other finally balances gravity's inward force, and the contraction ceases. The star has become a *neutron star*.

Neutron stars are very unusual objects. Their density is the same as the density in the nucleus of atoms: about 100 billion kilograms (about 100 million tons) per cubic centimeter. Atoms themselves are mostly empty space, with electrons whizzing around the collection of protons and neutrons called the nucleus, where nearly all the mass of the atom is found. Hence an entire star with the density of an atomic nucleus is incredibly small. A star twice the mass of our Sun would be compressed into a sphere no more than about 5 miles in diameter. There are literally thousands of asteroids in our solar system larger than that, but all of them have a density, like the planets, of just a few grams per cubic centimeter.

Since neutron stars are so massive, yet so small, the force of gravity on their surfaces is very strong—strong enough, according to Einstein's theory of general relativity, to bend light around them. This can cause some unusual optical effects. One example is illustrated at the site **http://www.tat.physik.uni-tuebingen.de/ENGL/ Forschung/compphys/movies/double.html**, which illustrates the hypothetical situation of Earth orbiting a neutron star. As Earth goes behind the star, the light coming from Earth bends strongly around the star, creating double images and even a doughnut-shaped image of Earth that makes it appear as if Earth is a ring around the star (see Figure 9.1).

There is more general information about neutron stars online. The astronomy department at the University of Chicago has an introduction to neutron stars that discusses their formation and physics, at **http://astro.uchicago.edu/home/web/miller/nstar.html**.

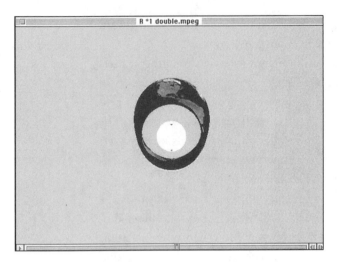

FIGURE 9.1
The Earth appears to form a ring around a neutron star in this simulation, because of the light-bending gravity of the star

Because neutron stars are so small and no longer generate energy from nuclear fusion, they are very difficult to observe from Earth, and essentially impossible at normal visible wavelengths. There are some illustrations available, however, of what neutron stars should look like, especially those that are in binary systems with other stars. Some examples are at **http://www.astroscu. unam.mx/neutrones/NS-Picture/NS-Picture.html**, which has some illustrations of known or theorized neutron stars, including one in the star system Centaurus X-3 (see Figure 9.2). (Read the next section for more about this star.) There is another illustration of a theorized neutron star that's part of a binary star system at **http://guinan.gsfc.nasa.gov/Images/pretty_pictures_general.html** (see Figure 9.3). There have even been simulations to see what would happen if two neutron stars collided (an unlikely but not impossible event). The results of the studies, including some images from a supercomputer-generated animation, are online at **http://141.142.3.132/SCMS/DigLib/text/astro/3D-Neutron-Star-Evans.html** (see Figure 9.4).

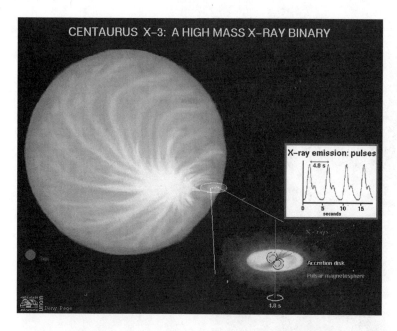

FIGURE 9.2
An illustration of the Centaurus X-3 system, which features a neutron star and a super-giant star in a binary system

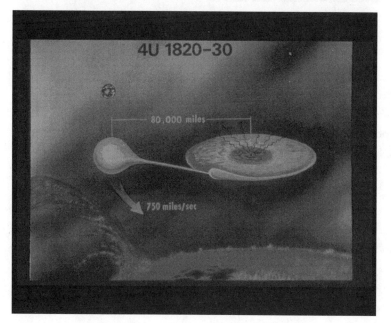

FIGURE 9.3
An illustration of a binary star system that includes a neutron star, that has captured some of the matter from its companion star into an accretion disk

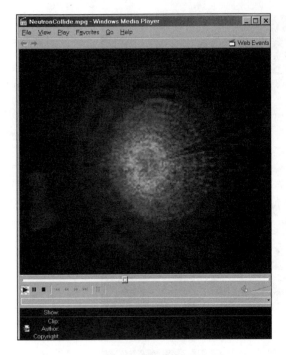

FIGURE 9.4
A frame from a supercomputer simulation of two neutron stars colliding

Pulsars

As exotic as neutron stars are, they are nearly impossible to detect because of their tiny size and the fact that they no longer generate energy via fusion, like normal stars. Neutron stars were first proposed in the 1930s but the idea seemed so outrageous that, with the lack of observational evidence, the concept was dismissed for decades.

The concept would gain new life, though, after the fall of 1967. At that time a team of Cambridge University astronomers, including graduate student Jocelyn Bell, were working on a new, highly sensitive radio telescope system at Jodrell Bank, England. Analyzing data the telescope collected, Bell found that one, unknown object was emitting radio beeps every one and one-third seconds, with extreme precision and without ever missing a beep.

No one at the time knew what could generate beeps with such regularity (the time between beeps was accurate to millionths of a second). Astronomers even seriously considered the possibility that this object, dubbed a *pulsar*, was some kind of beacon set up by an extraterrestrial intelligence. That idea, though, fell by the wayside as more pulsars, with different frequencies, were discovered. The question of the nature and origin of pulsars remained open, though.

Then a pulsar was discovered inside the Crab nebula. The Crab nebula had been accepted for some time to be a remnant of a supernova explosion. It seemed likely that the pulsar was tied somehow to the supernova explosion. The frequency of the beeps indicated the object was spinning around itself at the rate of about 30 times per second. Only one hypothesized object could do that and not fall apart: a neutron star.

It seems likely today that most neutron stars are pulsars. Neutron stars spin at high speeds, from the conversation of angular momentum: The smaller the radius of an object is, the faster it will spin. Just as an ice skater spins faster when she pulls in her arms, a star will spin faster when it changes from a massive supergiant to a tiny neutron star. This rapid rotation interacts with the star's intense magnetic field, generating electricity as the magnetic field moves around the star's rotation axis, usually at some angle to the axis. The electricity affects any protons and electrons on the surface, ejecting them through the star's polar regions. These particles emit energy as they accelerate through the magnetic field in a pair of beams, one from each pole. That energy is the radio pulse that telescopes record, and it's detectable only when the pole comes within Earth's line of sight; hence the sequence of regularly spaced pulses as the beam from the pulsar sweeps past Earth.

Another type of pulsar was discovered a few years later. Scientists analyzing X-ray observations of the universe from an unmanned satellite found a set of X-ray pulses coming from the constellation Centaurus. At first it seemed this source, named Centaurus X-3, was exactly the same kind of rapidly, regularly repeating source as the original pulsars, but visible in X-rays instead of radio waves. However, for 12 hours once every three months, the signal disappeared: a clue that some other object, like a star coexisting with the pulsar in a binary system, was blocking the signal.

We now recognize this interruption of X-ray pulses to be the signal of a neutron star in a binary system with an ordinary star. The neutron star's powerful gravity draws off gas from the outer layers of its companion star. The gas circles the neutron star in a lens of matter known as an *accretion disk,* slowly spiraling in and heating up. The magnetic field channels the gas to its poles. When the gas strikes the star's surface, it releases intense energy in the form of X-rays emitted through the poles. Just as with the poles of an ordinary pulsar, we see these X-rays as the beams sweeps across Earth in a rapid, regular manner.

The are two good introductions to pulsars online. The Royal Greenwich Observatory has a short but informative outline of our knowledge of pulsars, one in a series of information leaflets. The leaflet is on the Web at **http://www.ast.cam.ac.uk/pubinfo/leaflets/ pulsars/pulsars.html**. Since pulsars were discovered at Jodrell Bank radio observatory, it should be no surprise that one of the best introductions to pulsars is available there. This site includes a thorough description of pulsars, written for a graduate class but still very useful for general readers. It's online at **http://www.jb.man. ac.uk/~pulsar/tutorial/tut/tut.html**.

In addition, the program Galaxy Guide for Windows has some brief explanations of pulsars and neutron stars, including a description of the connection of the pulsar in the Crab nebula to neutron stars (see Figure 9.5).

Because pulsars usually shine in radio or X-ray wavelengths, they are hard to observe, and there are few good images of them. One good image was taken at gamma-ray wavelengths (gamma rays have shorter wavelengths than X-rays and thus are even more energetic) of the Crab pulsar and another nearby pulsar, at **http://cossc.gsfc.nasa.gov/cossc/outreach/pictures/galcen.html** (see Figure 9.6). This image was taken with the Compton Gamma Ray Observatory, an orbiting telescope that observes gamma rays, which are absorbed by the atmosphere and thus are not visible on

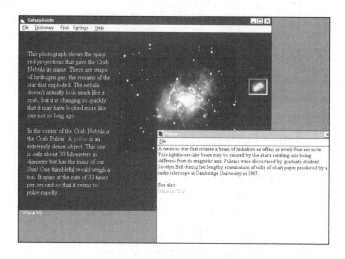

FIGURE 9.5
Information about pulsars, in particular the one found in the Crab nebula, from the program Galaxy Guide for Windows

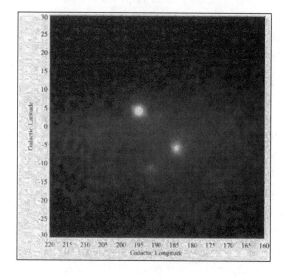

FIGURE 9.6
A gamma ray image of two pulsars, including the Crab pulsar, from the Compton Gamma Ray Observatory

the Earth's surface. In the image the pulsars look like ordinary stars, but in reality they are pulsing at many times a second.

Black Holes

Neutron stars are supported from further collapse by the pressure neutrons exert on each other in the densely packed interior. This works for stars of masses up to about three times the Sun's. However, for even more massive stellar remnants left behind by supernova explosions, the force of gravity becomes so strong that not even the neutrons can resist it. What keeps the star from collapsing further? Nothing. No known force in the universe can keep the star from collapsing into an infinitely small, infinitely dense point.

Our knowledge of physics is not advanced enough yet to say what form matter takes in such cases, but we do know some things about the characteristics of such bodies. One key characteristic is that the escape velocity—the speed an object requires to escape their gravitational pull—exceeds the speed of light. Since nothing in the universe can travel faster than light, nothing can escape the pull of one of these objects, not even the light they emit. These objects are called black holes.

Although we can't see the surface of a black hole, we can define an effective surface, the distance from the center of the hole at which the escape velocity is equal to the speed of light. This is called the *event horizon,* and the distance to this from the center of the black hole is called the *Schwarzschild radius,* after the early-twentieth-century astrophysicist Karl Schwarzschild, one of the first people to understand the concept of black holes. There is no physical surface at the event horizon; it marks the spot where the black hole, and anything within the Schwarzschild radius, is cut off from the rest of the universe. Anything that passes the event horizon toward the center of the black hole, even light, can never escape.

There are several good sources of information online about black holes. The Royal Greenwich Observatory has another in its series of information leaflets at **http://www.ast.cam.ac.uk/pubinfo/leaflets/blackholes/blackholes.html**, which goes over the basics of black holes. Beyond the Event Horizon at **http://www.astronomical.org/astbook/blkhole.html** is another good introduction. French author

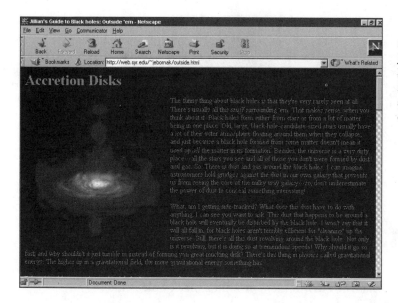

J.P. Luminet has a full chapter from his 1992 book *Black Holes* available for free online at **http://darc.obspm.fr/~luminet/chap10.html**.
The Black Holes FAQ (Frequently Asked Questions), by UC Berkeley astrophysicist Ted Bunn, provides a detailed set of answers to
common questions about black holes at **http://physics7.
berkeley.edu/BHfaq.html**. Jillian's Guide to Black Holes, **http://web.
syr.edu/~jebornak/blackholes.html**, is an illustrated look at the
topic (Figure 9.7). Offline, the program The Universe: From
Quarks to the Cosmos, from the Scientific American Library, has
some good information and animations of black hole simulations,
including explanations by astrophysicist Martin Rees (see Figure 9.8).

One of black holes' most interesting aspects is their ability to
act as a laboratory of sorts for studies of relativity. Their gravitational force is so intense it warps the space-time continuum around
it, creating effects not seen anywhere else. Although you would
have to travel to (and even in) a black hole to see some of these
effects, computers can simulate them to give you a better (and

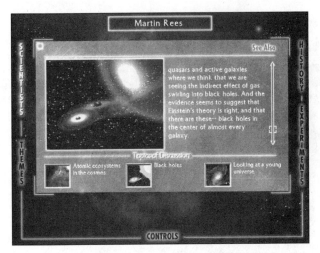

FIGURE 9.8
Astrophysicist Martin Rees
explains the nature of black holes
with the aid of computer-generated animations in the program
The Universe from Scientific
American Library

safer!) idea of what the environment around a black hole is like.
For example, Virtual Trips to Black Holes and Neutron Stars,
at **http://antwrp.gsfc.nasa.gov/htmltest/rjn_bht.html**, features
computer-generated movies in MPEG format that illustrate some
of the effects of flying close—even too close—to a black hole.
Among these effects is the bending of light around a black hole as
one approaches it. In a frame from one of those movies, two sets of
stars are visible; they are the same stars, but their light is bending
around the black hole in two directions, creating double images
on either side of the black hole as the observer approaches it (see
Figure 9.9).

To get a better understanding of how a black hole can warp
space and bend time, visit **http://math1.uibk.ac.at/~werner/bh/**. This
site features ray-tracing simulations of the effects of the black hole's
strong gravity field on a make-believe background (see Figure 9.10).
By comparing the left image, which uses "normal" Newtonian
physics that don't include the bending of light or the warping of
space-time, with the right one, which takes relativity into account,
you can see what sort of odd effects black holes can create.

To get an idea of what it would be like to fall into a black hole,
check out the series of animated GIF movies at **http://casa.colorado.
edu/~ajsh/schw.shtml**. These show simple computer models of what

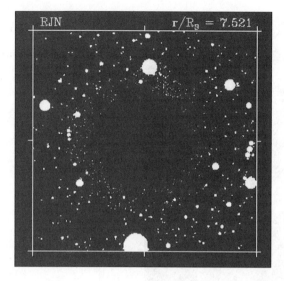

FIGURE 9.9
A frame from an MPEG movie of an approach to a hypothetical black hole—notice how the star patterns are repeated on opposite sides of the hole

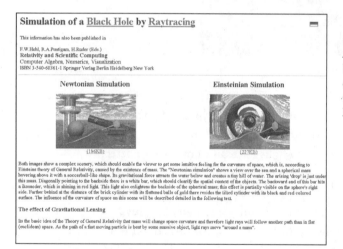

FIGURE 9.10
A ray-tracing simulation of a black hole's light-bending effects, assuming the sphere in the center of the image acts like one

you would see as you approached, orbited, and then passed through the event horizon of a black hole. University of Arizona astrophysicist Sam Hart has developed more-sophisticated models that show the form black holes take at **http://physics.arizona.edu/~hart/bh/** (see Figure 9.11). The National Center for Supercomputing Applications features a number of movies that show advanced models of black

145

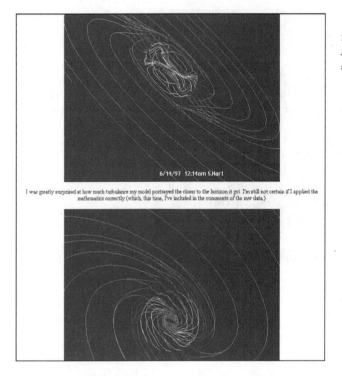

FIGURE 9.11
A model of a black hole and its accretion disk

holes and their interactions with other objects, computed using advanced high-speed supercomputers. This archive of QuickTime movies is online at **http://jean-luc.ncsa.uiuc.edu/Movies/**.

Since black holes don't allow light to escape, how can we detect their existence, short of trying to observe some of the unusual relativistic effects they create? There are several indirect methods. Astronomers detected the first black hole candidate, Cygnus X-1, when they found an intense source of X-rays that turned out to be one component of a binary star system with a supergiant star. The unseen companion was large enough—with seven times the Sun's mass—that it could only be a black hole. The X-rays came from gas pulled away from the supergiant by the black hole. This gas formed an accretion disk around the black hole, slowly spiraling in, heating up in the process and emitting X-rays before it was lost forever to the black hole.

Thanks to the Hubble Space Telescope's advanced resolving power, astronomers have found new ways to detect the existence of black holes indirectly. They found one candidate in the center of galaxy M87 by measuring how fast matter was going around the galaxy's center. At the speeds measured, the object in the center had to be extremely massive, but also very small. A black hole was the only object that could contain the required amount of mass— about 3 billion times the Sun's—in an area no larger than the Solar System. Such supermassive black holes are probably not created by a supernova explosion but by the collection of huge amounts of gas during the black hole's formation. An HST image of the showing a jet of high-speed gas streaming away from it (a process which may be related to the black hole in the galaxy's core) is part of a collection of black hole images from the Astronomy Picture of the Day archive at the Goddard Space Flight Center, **http://antwrp.gsfc. nasa.gov/apod/lib/black_holes.html** (see Figure 9.12).

A more sophisticated version of this method of finding black holes is now available to astronomers, thanks to a new instrument on the Hubble Space Telescope. The Space Telescope Imaging Spectrograph (STIS) was added to Hubble during a shuttle mission in February 1997. This sophisticated instrument gives astronomers the ability to quickly find variations in the velocity of stars going around galaxies. One of the first uses of this instrument was in galaxy M84. In a single observing session with the new instrument, astronomers recorded high rotation velocities around the core of the galaxy: evidence that a black hole also lurked in the center of this galaxy (see Figure 9.13). More information about this discovery is online at **http://oposite.stsci.edu/pubinfo/PR/97/12.html**. More-recent observations with Hubble have uncovered evidence for a black hole in another galaxy, NGC 4151, a Seyfert active galaxy. This spiral galaxy has a bright, starlike nucleus and other characteristics that indicate it may be home to a black hole, quasar, or other powerful object. The Space Telescope Science Institute's Web site at

FIGURE 9.12
An image of the center of galaxy M87, which includes a jet and a likely black hole at its center

http://oposite.stsci.edu/pubinfo/PR/97/18.html offers more information about this object (see Figure 9.14). There's now evidence for a supermassive black hole in the center of our own galaxy, according to research performed at UCLA and discussed at **http://www. uclanews.ucla.edu/Docs/HLSW379.html.**

While neutron stars, pulsars, and black holes are some of the most exotic known objects in the universe, they are just a small part of much larger groupings of stars. Galaxies can contain hundreds of millions of ordinary stars, as well as many neutron stars, pulsars, and black holes. In the next chapter we'll take a step back from individual stars, both ordinary and extraordinary, and examine these giant collections, from our very own Milky Way to distant groups of diverse galaxies.

FIGURE 9.13

An image of spectrographic data from the galaxy M84: the S-shape in the middle indicates high velocities around the center, and hence is evidence for the existence of a supermassive black hole

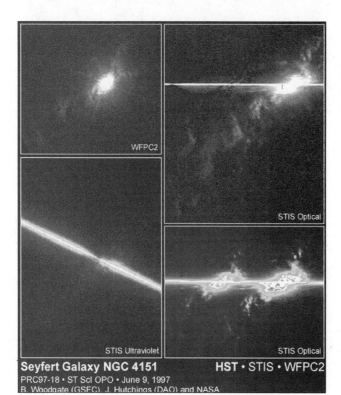

FIGURE 9.14

A collection of images from the Seyfert galaxy, NGC 4151, which shows evidence of a black hole in its center

10

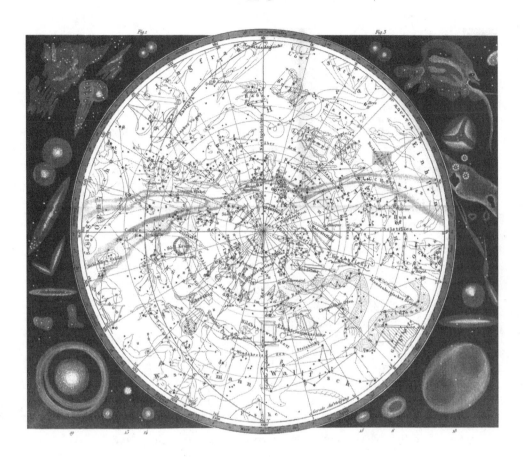

GALAXIES

G O OUT ON A CLEAR, DARK NIGHT, away from the influence of city lights, and look up. Odds are you'll see a distinctive band of light crossing the sky, composed of thousands of tiny stars too small and faint to resolve as individual stars. This feature, called the Milky Way, has been known since ancient times, but only in the modern era do we recognize the Milky Way to be a giant collection of stars—100 billion or more—which our Sun calls home.

If you're out on some late fall or winter evening, turn west. Look in the direction of the constellation Andromeda and you'll see, about halfway between the horizon and zenith, near the two "legs" of the constellation, a faint, fuzzy star. Examine this through a pair of binoculars and a distinctive spiral pattern will appear. This is the Andromeda galaxy, similar to our own and our celestial neighbor, although at two million light years away it is the most distant object visible to the naked eye.

Galaxies are collections of stars and come in a wide variety of shapes and sizes: spirals, disks, and other odd shapes. In this chapter we'll explore what galaxies are, starting with our home galaxy, and point out useful computer resources that will help you learn more about these key components of the universe.

The Milky Way

T he Milky Way has been a prominent feature in the night sky since ancient times, but not until Galileo turned his telescope to it was it clear that it was composed of countless tiny, dim stars. Eighteenth-century astronomers noted that this band of stars wraps all the way around the sky, and concluded that the Milky Way was a giant disc of stars including the Sun. In the 1780s famed astronomer William Herschel tried to find our place in the Milky Way by counting the number of stars in hundreds of different regions of the sky. He concluded that the density of stars doesn't change, and thus we were at the center of the Galaxy: a conclusion that—through no fault of his—turned out to be wrong.

Only in this century have we begun to understand the Milky Way's true nature. In the early part of the century, astronomer Harlow Shapley measured the distance to a number of globular clusters—small, spherical collections of stars—by measuring the period of a type of variable stars caller RR Lyrae stars, which behave similarly to Cepheid variables (see Chapter 7). He noticed that most globular clusters are located around one constellation, Sagittarius, and that they appear to gather around some point in that constellation. Since the clusters would be attracted to the Galaxy's center of mass, it was clear that the center was not the Sun, but that point about 30,000 light-years away from us.

What threw off Herschel's observations was the presence of interstellar dust, which dims light that passes through it. The stars he saw were closer than he thought; their light was dimmed, and light from more-distant stars was totally obscured by dust. Fortunately, radio wavelengths are largely immune from the effects of interstellar dust, and provide a much better view of the Galaxy. Astronomers using radio telescopes have found the disk of the Galaxy to be 100,000 light-years across but only about 2,000 light-years thick. Radio astronomers have also detected the presence of spiral arms, bands of stars that wrap around the core of the Galaxy in long, sweeping arcs. There's also some evidence that a central bar of stars goes through the galactic core, but that is still being debated.

We often think of galaxies as being vast objects deep in space, beyond the means of the typical amateur astronomer to observe without optical aid. There are two galaxies, however, that observers in the Northern Hemisphere can observe readily at various times of the year. The first, and most obvious, is the Milky Way—our own galaxy; the second is the great Andromeda galaxy, which is a naked eye object in even moderately dark skies. Simple binoculars can add even more to the enjoyment of observing these vast collections of stars.

Observers in the Southern Hemisphere are fortunate in being able to add two more naked eye galaxies—the Large and Small Magellanic Clouds, companions to our parent galaxy.

Even a modest telescope brings a plethora of galaxies into the range of visibility for the amateur astronomer. Observing the different shapes, sizes, angles, and details of these vast seas of stars can provide endless hours of enjoyment. Modest instruments can even observe the results of galaxies in collision, such as the beautiful "Whirlpool" galaxy, M51. It's hard for me to look through an eyepiece at one of these vast assemblages of stars without wondering if there's not someone on a world circling one of those suns who's observing the night sky as I am. Sometimes observing is more than just looking. . . .—R.L.

Radio astronomy has also shown that the Milky Way rotates. By measuring the velocities of stars closer to and farther from the center of the Galaxy and comparing those speeds to the random motion of other nearby stars and globular clusters, astronomers found that the Sun is moving around the center of the Milky Way at about 230 kilometers a second, or over 500,000 miles per hour. By comparison, Earth moves around the Sun at a speed of only 30 kilometers per second, or 67,000 miles per hour. Despite that tremendous speed, the Sun still needs nearly 250 million years to complete a single orbit around the Galaxy's center. That one figure gives an idea of just how large the Milky Way truly is.

There are several resources of information about the Milky Way online. The Perth Observatory in Australia has a basic introduction at **http://www.wa.gov.au/perthobs/hpc2mil.htm**. A somewhat more detailed introduction is provided by Cambridge University at **http://www.damtp.cam.ac.uk/user/gr/public/gal_milky.html**. Another site, **http://cgibin1.erols.com/arendt/Galaxy/mw.html**, discusses the shape of the Galaxy, including descriptions of computer models used to show how it evolved into its current shape over billions of years (see Figure 10.1).

There is also some information available offline about the Milky Way. Redshift, available for Windows and Macintosh, has an animated lesson on the Milky Way including descriptions of the size and shape of the Galaxy and the Sun's place in it (see Figure 10.2). The program Galaxy Guide for Windows has some basic information on the Milky Way galaxy, including a discussion of Baade's Window, a tiny portion of the sky in the constellation Sagittarius that astronomer Walter Baade studied intensely. He was able to see through this window past the center of the Milky Way to the stars just on the other side, which allowed him to determine the distance to the Galaxy's center (see Figure 10.3).

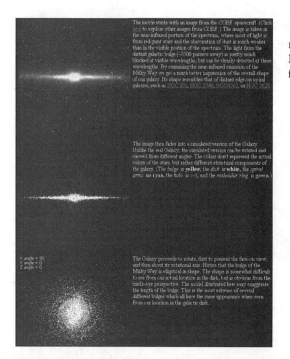

FIGURE 10.1

Images of a computer model describing the formation of the Galaxy

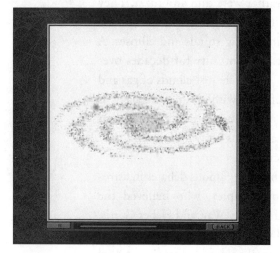

FIGURE 10.2

An animation showing the shape of the Milky Way galaxy, from the program Redshift

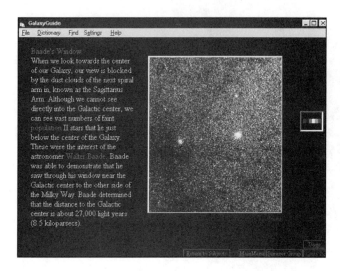

Baade's Window
When we look towards the center of our Galaxy, our view is blocked by the dust clouds of the next spiral arm in, known as the Sagittarius Arm. Although we cannot see directly into the Galactic center, we can see vast numbers of faint population II stars that lie just below the center of the Galaxy. These were the interest of the astronomer Walter Baade. Baade was able to demonstrate that he saw through his window near the Galactic center to the other side of the Milky Way. Baade determined that the distance to the Galactic center is about 27,000 light years (8.5 kiloparsecs).

FIGURE 10.3
A screen from the program
Galaxy Guide describing
Baade's Window in Sagittarius

Other Galaxies

As astronomers studied the sky with more powerful telescopes, new objects came into view. These objects did not have the pinpoint sharpness of stars, but rather were dim and fuzzy. They were dubbed *nebulae* from the Latin word for clouds. Some of these nebulae had distinctive shapes, including spirals and ellipses. A debate ranged in the astronomical community for decades over what they were. Some argued that they were just clouds of gas and dust, with perhaps small clusters of stars, within our own galaxy, as some other nebulae had turned out to be. Others claimed that these nebulae were very distant, and thus very large: *island universes,* as one astronomer called them.

This discussion led to one of the most famous debates in astronomical history, between Harlow Shapley, who believed the nebulae were part of or very near the Milky Way, and Heber Curtis, a young astronomer who believed they were separate galaxies. The 1920 debate didn't produce a winner, but did generate additional debate on the subject and led to a resolution of the argument by one of the most famous astronomers of the twentieth century.

Edwin Hubble, who was trained as a lawyer and worked as a schoolteacher before getting his Ph.D. in astronomy from the University of Chicago, went to work around 1920 at Mount Wilson Observatory, which was home at the time to the world's largest telescope, the 100-inch Hooker. In 1923 he used the Hooker telescope to take an image of the Andromeda nebula, as it was then known. It had a distinctive spiral shape and was considered a leading candidate to be a separate galaxy. In that image Hubble found a Cepheid variable star. (The relationship between the period and luminosity of these stars gives astronomers a way of finding distance.) Hubble tracked this and other Cepheid variables he saw in the nebula, and by late 1924 concluded that the nebula was very far from Earth: about 2 million light-years, using modern measurements. The debate was now over: many of the nebulae were very distant, and thus had to be very large. They were separate galaxies.

Since then astronomers have found that the universe is full of billions of galaxies, perhaps more galaxies than there are stars in our own Milky Way. Hubble developed a system to categorize galaxies based on their appearance, dividing them into four classes: spirals, barred spirals, ellipticals, and irregulars, which are galaxies with unusual and unique shapes. These classes are further subdivided, based on the tightness of the arms in spiral and barred spiral galaxies and the roundness of elliptical galaxies.

A wide range of resources on the Internet provide introductions to galaxies, including Science Backgrounder: An Introduction to Galaxies at **http://www.xtec.es/recursos/astronom/hst/galintro.htm**. Galaxies—A Popular Overview, at **http://www.astro.uu.se/~ns/ popgal.html**, describes galaxies in detail, including information on the history of our understanding of galaxies, their classification, and the relationship of galaxies to topics like dark matter and cosmology. The Royal Greenwich Observatory also has some basic information in its series of information leaflets at **http:// www.ast.cam.ac.uk/pubinfo/leaflets/galaxies/galaxies.html**.

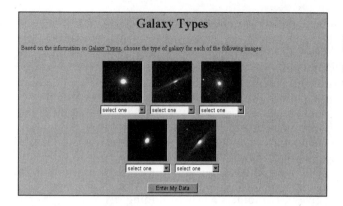

Test your knowledge of galaxy classification at Hands-On Universe

One of the best uses of the Web for teaching people about galaxies is Hands-On Universe at **http://hou.lbl.gov/ISE/new/galaxt** (see Figure 10.4). Here users can get a tutorial on the different types of galaxies, and then get quizzed on this knowledge by classifying galaxy images.

There are plenty of sites with images of galaxies as well. New Zealand's Carter Observatory includes a few images of galaxies in its introduction to the subject at **http://www.vuw.ac.nz/~bankst/ articles/galaxies.html**. At **http://www.eia.brad.ac.uk/btl/m4.html** you'll find descriptions of a number of major galaxies, such as the Andromeda galaxy and the Sombrero galaxy, so named because the spiral galaxy is viewed edge-on and has a large core, giving it the appearance of a Mexican hat. Cambridge University also has some images of galaxies to go with its descriptions at **http://www.amtp. cam.ac.uk/user/gr/public/gal_home.html**.

Larger and more specific collections of images are available. At **http://seds.lpl.arizona.edu/messier/** you'll find a listing of Messier objects, which are nebulae and other nonstellar objects cataloged by French astronomer Charles Messier in the late eighteenth and early nineteenth centuries as a way to keep him and other astronomers from confusing them with new comets. Many of these

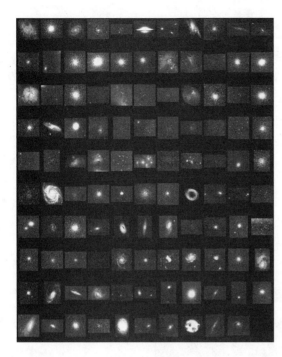

FIGURE 10.5
A collection of images of Messier objects, which includes a number of prominent galaxies

objects turned out to be galaxies, such as M31 (Andromeda). This site has a complete gallery of images of the Messier objects and information on which of the more than 100 Messier objects are galaxies (see Figure 10.5).

For a comprehensive database of galaxies (and other objects), check out the NASA/IPAC Extragalactic Database, or NED, at **http://nedwww.ipac.caltech.edu**. The database allows you to search by name and position, and retrieves information and any images of that galaxy. The database can also generate finder charts to locate the object's position relative to stars and other objects in the nght sky (Figure 10.6).

For a collection of galaxy images, turn to **http://astro.princeton.edu/~frei/galaxy_catalog.html** (see Figure 10.7). This set of images of 113 galaxies was obtained by a graduate student at Princeton University as part of his Ph.D. thesis. You can view the whole set or

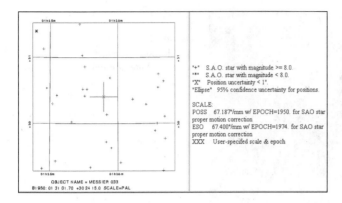

"+"   S.A.O. star with magnitude >= 8.0.
"*"   S.A.O. star with magnitude < 8.0.
"X"   Position uncertainty < 1".
"Ellipse" 95% confidence uncertainty for positions.

SCALE:
POSS   67.187"/mm w/ EPOCH=1950. for SAO star
proper motion correction
ESO   67.400"/mm w/ EPOCH=1974. for SAO star
proper motion correction
XXX   User-specifed scale & epoch

FIGURE 10.6
A finder chart for the galaxy M33 generated by NED's Sky-Plot routine

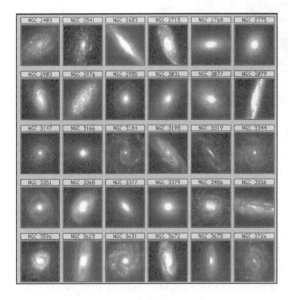

FIGURE 10.7
Part of the catalog of 113 images of spiral galaxies available from Princeton University

examine the catalog one galaxy at a time. The University of Alabama has a collection of images of *galaxy clusters*, groups of galaxies bound together by each other's gravity, at **http://crux.astr. ua.edu/white/mug/cluster/clusters.html**.

The Hubble Space Telescope has taken a number of good galaxy images. One of the most dramatic is the Hubble Deep Field, an

FIGURE 10.8
The Hubble Deep Field, an image of a small section of the sky that shows thousands of different galaxies of various types

image of a very small portion of the sky created by combining images taken over a ten-day period (see Figure 10.8). This shows a wide variety of galaxies in different shapes, sizes, and colors. One use of the Hubble Deep Field, as explained at **http://oposite. stsci.edu/pubinfo/PR/96/29.html**, is to find evidence of the building blocks for early galaxies by looking farther into the universe (and thus at an earlier stage).

Other good Hubble images include those at **http://oposite. stsci.edu/pubinfo/PR/96/31.html**, which shows a bright region in an otherwise dim galaxy where stars are being born. In another image, at **http://oposite.stsci.edu/pubinfo/PR/96/36.html**, Hubble sees unusual cometlike clouds form in the heart of the Cartwheel galaxy, an unusual galaxy that's likely the remnant of a collision between itself and another galaxy.

More exotic and unusual images are available. NASA's Goddard Space Flight Center has images taken at X-ray wavelengths using the ROSAT satellite, online at **http://heasarc.gsfc.nasa.gov/Images/ rosat/galaxies.html**. Images of clusters of galaxies taken with ROSAT are also online at **http://heasarc.gsfc.nasa.gov/Images/**

rosat/clusters_galaxies.html. A sample of images of unusual galaxies is Arp's Catalog of Peculiar Galaxies at **http://members.aol.com/arpgalaxy/index.html** (see Figure 10.9). This site provides information on a catalog of several hundred unusual galaxies that astronomer Halton Arp compiled in the 1960s; it offers images, descriptions of the galaxies, and Arp's catalog. Also, the University of Alabama has collections of images of active galaxies that may be homes to black holes, quasars, or other energetic objects, at **http://crux.astr.ua.edu/active2.html**, and images of galaxy pairs, including galaxies interacting or even colliding, at **http://crux.astr.ua.edu/pairs2.html** (see Figure 10.10).

A Gallery of Remarkable Arp Views

Arp 278, NGC 7253

Arp 188, UGC 10214

Arp 222, NGC 7727

All Photos by Al Kelly, 1995
CB245 CCD, most on 32" f/4 Newtonian
False color by Dennis Webb

Arp 319, Stephan's Quintet

Arp 85, Messier 51

Arp 244, NGC 4038 & 4039

FIGURE 10.9
A gallery of some of the most interesting entries in Arp's Catalog of Peculiar Galaxies

FIGURE 10.10
An image of two interacting galaxies, UGC 2942 and 2943

Offline, several programs have information about and images of galaxies. Discovery Astronomy for Windows has a section devoted to galaxies, with information about the Andromeda galaxy, the Virgo cluster of galaxies, and a good selection of galaxy images. The Universe, by Scientific American Library for Macintosh and Windows, features interviews with astronomers who do research related to galaxies. They discuss the formation of galaxies and the large-scale structure of galaxies spread across the universe. As the name suggests, Galaxy Guide for Windows offers a good introduction to galaxies and a number of good images.

Observing Galaxies

If this discussion has interested you, the next step you might want to take is to go out and look for galaxies in the night sky. Only the very brightest one, the Andromeda galaxy, is visible to the naked eye. However, even with a basic pair of binoculars, many more come into view. To find a particular galaxy, though, you need to know where to look.

If all you need is a list of galaxies and their coordinates, check out **http://www.seds.org/messier/xtra/supp/rasc-g-n.html**. This is a simple list of some of the brightest galaxies visible, with coordinates in right ascension and declination. If you're familiar with the positions of objects in the night sky and know, or know how to find out, where an object with a specific right ascension and declination would be, this list alone may be sufficient.

However, if you need star maps and more detailed information to find galaxies, particularly dimmer ones, you will want to turn to one of several good planetarium programs with information about their location and visibility. Redshift includes many galaxies in its database of objects. You can select a galaxy already visible in the sky view, or find a specific galaxy. Once you've selected one, you'll get its location and visibility information. For some major galaxies, the program offers images and other background information, as in the case of the spiral galaxy M81 (see Figure 10.11).

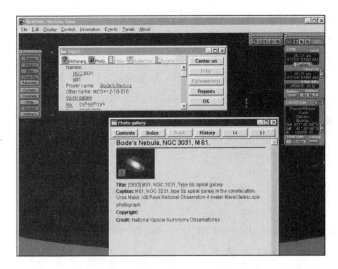

FIGURE 10.11
Information on the galaxy M81 from the program Redshift

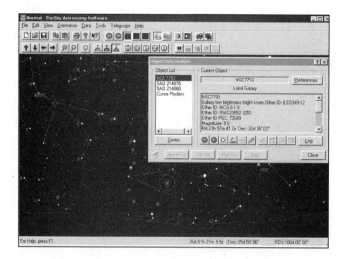

FIGURE 10.12

Information provided by the program The Sky for a typical galaxy, in this case NGC 7793

Distant Suns for Macintosh and Windows also provides access to a large database of galaxies, and includes images for many of them, even for some of the lesser-known, dimmer galaxies. There's also a find utility to locate a specific galaxy, although sometimes it can be difficult to figure out which of the many galaxies displayed on the screen is the one in question.

If you're not interested in pretty pictures but want detailed information on bright and dim galaxies, you may want to consider The Sky for Windows, which offers detailed information on even the dimmest galaxies, like NGC 7793 (see Figure 10.12). You also have the option of linking the program to a computer-controlled telescope, so you can direct your telescope to a galaxy you point out in the program and spend more time actually observing the object instead of manually trying to find the galaxy with the telescope.

If your primary observing interest lies with galaxies and other deep sky objects, you may want to consider a specialized program like NGCView for Windows, by Rainman Software. This program is a deep-sky observational planning tool; it allows the user to select the objects they want to observe, generate lists of planned observations, and then log those observations. The planning aspect of the program gives you detailed information on a selected galaxy, including its location, brightness, and rise and set times. You can use a star chart it generates to help find the object, and an altitude plot shows how high in the sky it is during each part of the night and where the Moon is, so you can avoid making observations when the Moon is nearby and bright (see Figure 10.13).

Many of the nebulae cataloged in the eighteenth and nineteenth centuries have turned out to be separate galaxies, some like our own, and others very different. However, many other nebulae are not galaxies at all, but clusters of stars or clouds of dust and gas. The next chapter will focus on these other nebulae and star clusters.

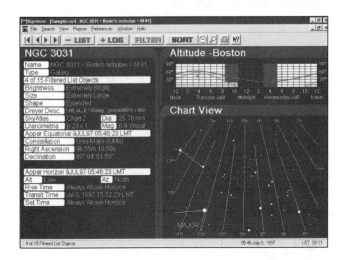

FIGURE 10.13
Detailed viewing information for the galaxy M81 provided by NGCView

11

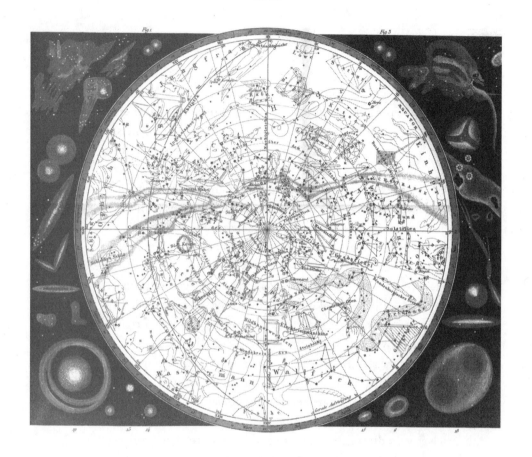

NEBULÆ AND CLUSTERS

A S TELESCOPES ENTERED INTO USE in the seventeenth century, astronomers discovered that there was more in the night sky than just stars and planets. Astronomers found an increasing number of "extended" objects—objects that appeared larger than stars in any telescope—of various shapes and colors. Many of these had a cloudlike appearance and became known as nebulae, Latin for clouds. Others appeared to be dense collections of stars, which became known as clusters. As telescopes become more powerful, more nebulae and clusters were discovered.

Today we recognize that nebulae are clouds of gas and dust. Some are cold, dark clouds of interstellar dust that obscure the light from stars and galaxies behind them, while others shine in bright colors created by their own light or the reflected light of other objects. Clusters are collections of stars; some within our own galaxy that are fairly small and open, while others found around our own and other galaxies are dense and globular.

While nebulae and clusters share few characteristics and are themselves very diverse, astronomers often refer to them collectively as *deep sky objects*. They are mixed together in catalogs with galaxies, which were originally identified as a type of nebula but are now recognized as a much different class of object. The wide range of nebulae shapes and colors, which include some strikingly beautiful forms, make them popular and sometimes challenging targets for amateur astronomers. In this chapter we'll look at these deep sky objects and find some software and online resources for learning more about them.

History

Ancient astronomers knew of a few nebulae and clusters long before the telescope entered the picture. Certain large, bright clusters of stars such as the Pleiades have been known since ancient times, and Greek and Persian astronomers found a few nebulous stars that later turned out to be star clusters. One supposed nebula was probably known to ancient astronomers—the Andromeda nebula, which was not recognized as a galaxy until the twentieth century. The first discovery of a true nebula of gas and dust did not occur until the era of telescopes in astronomy began in the early seventeenth century, when French astronomer Nicholas-Claude Fabri de Peiresc found M42, the Orion nebula. Other astronomers occasionally discovered new nebulae and clusters throughout the seventeenth and eighteenth centuries.

The discoveries of deep sky objects got a boost from the cataloging efforts of Charles Messier. The French astronomer assembled a catalog of over 100 deep sky objects, including nebulae and clusters (many of those nebulae were later found to be galaxies), and published it in 1781. His goal was not so much to further the field of deep sky astronomy as simply to end the confusion among the different, smaller catalogs of such objects assembled by other astronomers. Messier's interest was in discovering new comets, and he wanted a catalog of known deep sky objects like nebulae, which one might confuse with a comet. His catalog is in use to this day: Objects that appear in the catalog are identified with the letter M and a number—for example, the nebula M42 mentioned above.

A contemporary of Messier, William Herschel, also made important nebula and cluster discoveries, assembling his giant catalog of 2,500 deep sky objects, including nebulae and clusters, by the early 1800s. He

Those new to observing through telescopes are often disappointed when first observing nebulae through their new instrument, expecting a bright image and the full blaze of glorious color he or she has seen in the famous astronomical photographs of these objects. In truth – they are faint, and unless one has a particularly large telescope, the color seen will likely be a faint green. Due to the human eye's particular sensitivity in the yellowish-green part of the spectrum and its inability to distinguish color in low light levels, this is a fact of life. There is detail, however, and sometimes lots of it. In addition to the greenish glow, I've seen traces of red in the Orion nebula (M42), a particularly bright object, but not in any other nebula in an amateur-size telescope.

A group of specialized filters exists, however, that can aid in observing detail in such objects. Often called nebula filters, these filters have special coatings that will both darken the sky and block out the effects of light pollution, while passing freely the specific bands of light for which they are designed. By increasing the contrast, these filters can be a tremendous aid in both observing and astrophotography, particularly in more urban areas. They typically screw into the back end of the eyepiece used for observing, although versions are available for cameras, and are made by a variety of manufacturers.—R.L.

also classified the nebulae and clusters by their appearance, although these descriptions are less useful today, since we better understand these objects. Herschel's son, John, traveled to South Africa in the 1830s and did a similar search of the southern skies, later assembling the "General Catalogue" of more than 5,000 deep sky objects.

For a more thorough, complete history of the discovery of deep sky objects, see an online Messier database at **http://www.seds.org/ messier/xtra/history/deepskyd.html**. This site has a detailed account of the discovery of nebulae and clusters from ancient times to the work of John Herschel, which brought the great era of deep sky discoveries to an end. This page includes links to information about the objects these astronomers discovered, as well as other reference works in this field.

Nebulae

Nebulae were once defined as any cloudlike object visible in the night sky, but as many of those turned out to be galaxies, we now consider nebulae to be clouds of dust and gas. This is still a very open definition; there is a wide range of nebulae, from dark, black objects that obscure starlight to those that shine a bright red or blue. Their cloudlike nature also means that nebulae come in almost any imaginable shape.

There are different methods of classifying nebulae. One way is by whether they emanate their own light, reflect light from another object, or absorb light. *Absorption,* or dark, nebulae are cold clouds of dust and gas that absorb the light they encounter; thus they appear as black clouds against a starry background. An example of an absorption nebula is the Horsehead nebula: the dark cloud, which has the appearance of a horse's head, is closer to us than the brighter nebula and stars behind it, and blocks them out, creating a dramatic contrast (see Figure 11.1). Absorption nebulae actually reemit the light they receive to maintain an equilibrium temperature, otherwise they would heat up indefinitely. However, they do

FIGURE 11.1
An image of the Horsehead Nebula, from the program Discover Astronomy for Windows

so at much longer wavelengths than visible light, and thus appear dark to us.

Emission nebulae, like absorption nebulae, absorb light from stars and reemit it at longer wavelengths. However, in the case of emission nebulae, the light is absorbed at ultraviolet wavelengths invisible to the human eye and emitted in visible light, usually red. This makes these objects appear as bright objects in the night sky. An excellent example is the Orion nebula: The clouds of gas and dust in it absorb ultraviolet light from new O and B class stars within the nebula, and then reemit the light at visible wavelengths. A good Hubble image of the Orion nebula, with some additional information about it, is at **http://oposite.stsci.edu/pubinfo/PR/95/45.html** (see Figure 11.2).

Reflection nebulae do not primarily absorb or emit light, but instead reflect light from stars within or near them. The dust grains in the nebula scatter starlight to give these nebulae their bright appearance; the diffuse nebula around the Pleiades are an example (see Figure 11.3). Since the shorter wavelengths of blue light scatter far better than longer wavelengths of light (that's the reason why the sky is blue), reflection nebulae are usually blue.

FIGURE 11.2
An image of the Orion Nebula, M42, from the Hubble Space Telescope

FIGURE 11.3
An image of the Pleiades cluster, with the hazy glow of its reflection nebula easily visible, from Redshift

Another way to classify nebulae is based on their shapes. One such class is the *planetary* nebulae, so named because when astronomers such as William Herschel first viewed them in telescopes they appeared similar to planets (Herschel was interested in both planets and nebulae, having discovered the planet Uranus in addition to the many nebulae he recorded). These objects have a disklike shape, and some even appear to have rings. Planetary nebulae form when a star such as the Sun reaches the end of its life and

Planetary Nebula NGC 7027 HST · WFPC2
PRC96-05 · ST ScI OPO · January 16, 1996 · H. Bond (ST ScI) and NASA

FIGURE 11.4
A Hubble Space Telescope view of the planetary nebula NGC 7027

ejects much of its mass as it collapses (see Chapter 8 for more information on the life cycle of stars like the Sun.) Planetary nebulae like NGC 7027 (see **http://oposite.stsci.edu/pubinfo/PR/96/05.html** for a good image of this nebula; Figure 11.4), are generally emission nebulae; they absorb light from the hot, dying star and reemit the light at visible wavelengths. The opposite of planetary nebulae are *diffuse* nebulae, which can take on a wide variety of shapes. Most nebulae, particularly remnants from supernova explosions, such as the Crab nebula, are diffuse.

Planetary nebulae are among the most studied, due to their connection to the death of stars. A Planetary Nebulae Sampler, at **http://www.noao.edu/jacoby/pn_gallery.html**, provides a gallery of images of different planetary nebulae, with some discussion about each one shown. A larger list of planetary nebulae, complete with images, is at **http://www.ozemail.com.au/~mhorn/pneb.html**. A more complete list of planetary nebulae can be obtained from a database at **http://ast2.uibk.ac.at/**. This site allows you to search for planetary nebulae based on name, position in the sky, and more. The database includes detailed information and images of the nebulae, although some of the information is advanced enough to be truly useful only to a professional or advanced amateur astronomer.

Gaseous Pillars · M16   HST · WFPC2
PRC95-44a · ST ScI OPO · November 2, 1995
J. Hester and P. Scowen (AZ State Univ.), NASA

FIGURE 11.5
The dramatic image of the fingers of
dust in the nebula M16, from the Hubble
Space Telescope

While some nebulae, like the Crab nebula, are associated with
supernovae and the death of stars, nebulae are also known for being
the birthplace of stars. Some of them, particularly emission and
reflection nebulae, are stellar nurseries where dust and gas come
together to form stars. The intense light of newborn bright stars
makes the nebula shine with reflected or emitted light. One of the
best images of stellar birth activities in a nebula (and arguably one
of the most dramatic images ever taken with the Hubble Space
Telescope) is a detailed image of the Eagle nebula, M16, at
http://oposite.stsci.edu/pubinfo/PR/95/44.html (see Figure 11.5).
Long fingers of dust can be seen in the image, where gas globules
from which new stars would emerge have protected the dust from
ultraviolet light.

There are also some good general resources for nebula information and images. One of the best Web sites is The Web Nebulae at
http://www.seds.org/billa/twn/. This site has detailed information
on different types of nebulae, including images of all the varieties
discussed here. Some of the best images are from the Hubble Space

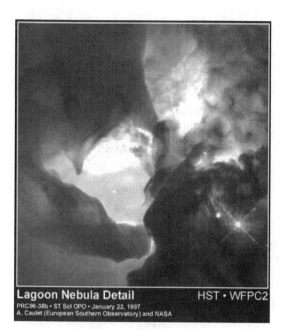

FIGURE 11.6
The Hubble's view of the Lagoon nebula, M8

Telescope, which has a list of publicly released nebula images, including objects like the Lagoon nebula (see Figure 11.6), at **http://oposite.stsci.edu/pubinfo/Subject.html#Nebulae.**

Offline, some software packages provide information about nebulae. Discover Astronomy, a program for Windows from Maris Multimedia, has some basic information about nebulae, along with some images. Galaxy Guide, another program for Windows by Parallax, also provides some basic information on the different types of nebulae, with images and illustrations.

Clusters

Clusters are collections of stars that not only appear together in the sky as seen from Earth, but are physically grouped together. Clusters are usually divided into two different groups: *open clusters* and *globular clusters*. Open clusters are collections of dozens to hundreds of usually young stars in an area no more than about 50 light-years across. These stars often formed from the same

175

I've always had a special fondness for these spherical conglomerations of stars, and even though the most magnificent representatives of the group are only visible to Southern Hemisphere observers, the northern skies also have many splendid examples. "Seeing," the transparency, clarity, and stability of the atmosphere, has much to do with observing these beautiful clusters successfully. I must admit, however, that one of the most memorable sights I've ever seen through my telescope is an observation I made of the cluster M13 in Hercules one late, warm, summer night.

The seeing was very poor that night, with lots of atmospheric turbulence, and M13 was relatively low in the sky. Being one of my favorites, I had to have a look. There it was, the famous cluster, with a multitude of small stars that scintillated with each wave of atmospheric turbulence. The seeing was so poor, indeed, that M13 appeared to be under a thin veil of rippling water, but it is this view of tiny twinkling stars that I carry with me as my favorite observation of M13. Had I not taken out my telescope on what I knew was a less than perfect night, I would have never made this observation.—R.L.

cloud of interstellar gas and dust, but are not gravitationally bound and will later go their separate ways. The Pleiades is a good example of an open cluster, and more information about the Pleiades is available at **http:// www.ras.ucalgary.ca/~gibson/pleiades/**.

Globular clusters are tightly bound, spherical collections of up to a million stars packed into an area no more than a few hundred light-years in diameter. While stars in our part of the Galaxy are usually separated by several light-years, stars in the heart of a globular cluster are separated by no more than a few light-*months!* These objects are not found in our galaxy, but surround it and other large galaxies in a spherical halo. Stars in globular clusters are much older than those in open clusters: some have been measured at up to 15 billion years old.

As it turns out, the age of globular clusters is a very important factor in finding the age of the universe. Other methods (which will be discussed in Chapter 13) have found that the universe is perhaps no more than 12 billion years old. However, the universe can't be older than the stars in the globular clusters. Since the age of globular clusters is based on our observations of the different types of stars there, which are mostly

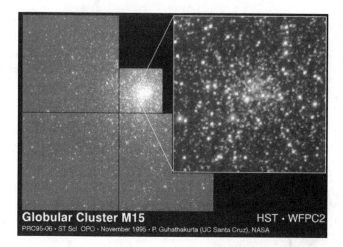

Globular Cluster M15   HST · WFPC2
PRC95-06 · ST ScI OPO · November 1995 · P. Guhathakurta (UC Santa Cruz), NASA

FIGURE 11.7
An image of the globular cluster M15, a typical globular cluster

long-lived red main-sequence stars—no brighter, shorter-lived yellow and blue main-sequence stars—we conclude that the clusters are very old. We believe we understand stellar evolution better than we understand the origin and age of the universe, so using the age of the globular clusters to determine the age of the universe is more reliable than other methods. But the universe's true age is still a mystery. A detailed catalog of globular clusters visible around our galaxy is available as a text file from **http://physun.physics.mcmaster.ca/GC/mwgc.dat**. A similar catalog, with more detailed discussion by astronomer Brian Skiff, is online at **http://www.ngcic.com/gctext.htm**. One of the best-known globular clusters, M15, has been imaged by Hubble; it appears at **http://oposite.stsci.edu/pubinfo/PR/95/06.html** (see Figure 11.7). This is a particularly good example of the densely packed, spherical globular clusters. While the globular clusters in this image are those around our own galaxy, they do exist around other galaxies, as seen in a Hubble image of one near the Andromeda galaxy at **http://oposite.stsci.edu/pubinfo/PR/96/11/A.html** (see Figure 11.8). Both Discover Astronomy and Galaxy Guide, described in the last section, provide information on open and globular clusters.

Globifular Cluster G1
in Galaxy M31
HST · WFPC2
PRC96-11 · ST ScI OPO · April 24, 1996
Michael Rich, Kenneth Mighell, and James D. Neill (Columbia University),
Wendy Freedman (Carnegie Observatories) and NASA

FIGURE 11.8
An image of globular cluster GC1, near the Andromeda galaxy, from the Hubble Space Telescope

Observing Deep Sky Objects

Deep sky objects like nebulae and clusters are favorite targets of amateur astronomers, given both their wide range of appearance and the challenge of finding these often dim objects. Trying to decide which objects to look for on a particular night, and how to find them, can be overwhelming; fortunately, a variety of online resources and software packages can help the amateur astronomer. A good starting place to get observing information is The Virtual Observatory, at **http://www.ifi.uio.no/~mikkels/astropics.html**. This well-designed site has information on a variety of deep sky objects, including details on all the objects in the Messier catalog, information on the larger New General Catalog (NGC), and other lists of deep sky objects. The Messier information is particularly good; you can search by Messier number, get information about the object, and select from a number of images (see Figure 11.9).

MESSIER 79

FIGURE 11.9
An image of the deep sky object M79, one of the many deep sky objects listed at the Virtual Observatory Web site

The NGC, which includes thousands of objects, is available in a searchable database at **http://www.seds.org/~spider/ngc/ngc.html**. Enter the number of the object in the NGC catalog, and the database will return some basic information, including its position in the sky and its brightness, as well as an image of the object and a link to additional information.

There are also lists of favorite deep sky objects available online. One of the best known is the Herschel 400, a set of 400 deep sky objects the famous astronomer compiled that are described as "challenging" to find. Many amateur astronomers set a goal of observing all 400 of these objects. You can find the Herschel 400 **http://www.seds.org/billa/herschel/** and **http://www.astronomical. org/alnotes/alher.html**. The Royal Astronomical Society of Canada has compiled another detailed list of "challenging" deep sky objects, available on the Web at **http://www.seds.org/messier/xtra/ similar/rasc-dsc.html**.

Many of the programs for finding and observing galaxies that were discussed in Chapter 10 are also useful for finding other deep-sky objects. The Sky for Windows and Starry Night Deluxe for Macintosh are both useful because they can be linked to certain telescopes. You can find and select a deep sky object in the computer, then have it direct the telescope to the appropriate location in the sky, relieving you of the work of finding the object manually. NGCView for Windows provides detailed information about deep sky objects. Redshift and Distant Suns, both for Macintosh and Windows, also provide information about nebulae and clusters, and both have good images of these deep sky objects (see Figure 11.10).

Nebulae and clusters are both fairly well-understood objects. But other deep sky objects in the universe are not at all understood, such as the mysterious sources of radio energy known as quasars. Astronomers are also trying to comprehend powerful, violent events taking place in active galaxies. These unusual, powerful objects will be the subject of the next chapter.

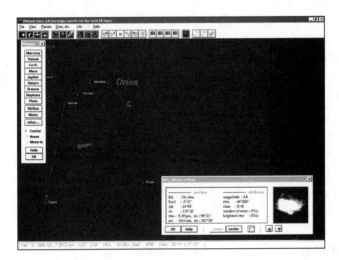

FIGURE 11.10
Information about the Orion nebula, including an image, available from the program Distant Suns

12

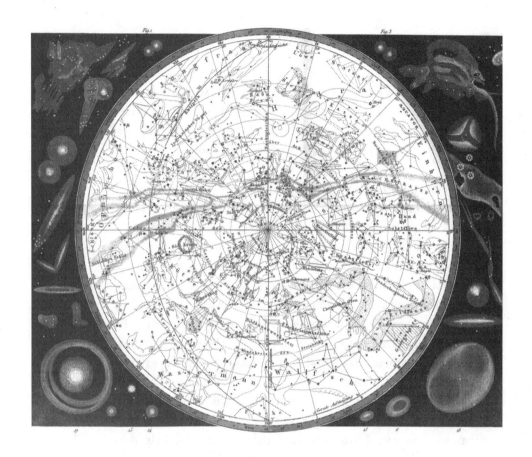

QUASARS AND OTHER ODDITIES

B Y NOW THE UNIVERSE MAY SEEM to make some sense. Planets orbit stars, which are found in galaxies and clusters. Stars are born, often along with solar systems; live out their lives; and die, sometimes quite spectacularly, depending on their mass. Clouds of dust and gas fill our galaxy and other galaxies as well, and are often tied to the birth and death of stars. Everything seems to be in order, at least as long as one is willing to accept the giant scales of mass, distance, and time that are commonplace in the universe.

However, there are objects and events that seem out of place when we accept this orderly view of the universe. Giant galaxies collide with one another. Other galaxies have intense bright cores, far too bright to consist of stars alone. Starlike objects appear to be billions of light-years away but are as bright as much closer objects, implying that these distant bodies are brighter than many galaxies put together. These objects, all of which astronomers have observed in recent decades, keep the universe from seeming too ordinary and too well understood. Although we now understand them better than when they were first discovered, their existence has helped us comprehend how amazing the universe can be.

In this chapter we take a look at some of these unusual objects—from colliding galaxies to unusually bright, or *active,* galaxies to quasars, the distant and incredibly bright sources of energy that have confounded astronomers for decades. We'll point out some good online and offline sources of information about these objects.

Colliding Galaxies

Space is incredibly vast, and although galaxies are large and numerous, most of the universe is still empty space. It thus seems unlikely that two galaxies would come close to one another, let alone come into contact. Yet we have seen several cases where galaxies are in the slow process of crossing each other's paths, with varying consequences.

What happens when a pair of galaxies come into contact depends largely on their sizes. If one galaxy is much more massive, the tidal forces it exerts on the smaller galaxy rip the smaller one apart. The individual stars in the smaller galaxy are not harmed, but are incorporated into the larger one. Astronomers have termed this kind of collision *galactic cannibalism.*

When the colliding galaxies are closer in size, they are more likely to merge into a single new galaxy. Computer simulations and observations of unusual galaxies show that these merged galaxies look like a very loose spiral galaxy, with a pair of arms stretching almost straight out on opposite sides of the center for tens of thousands of light-years. These arms are hurling thousands of stars into the cold void of intergalactic space. We haven't yet seen these stars in the space between galaxies, but later studies with the Hubble Space Telescope may discover these homeless survivors of galactic collisions.

Galaxy collisions can also take a form more like a hit-and-run accident. The Cartwheel galaxy takes its name from its bright nucleus surrounded by a ring of bright blue and white stars, with only tenuous gas and dust and a few stars between them (see Figure 12.1). Detailed observations with the Hubble Space Telescope (see **http://oposite.stsci.edu/pubinfo/jpeg/Cartwheel.jpg**) show that a second, smaller, irregular galaxy, at about the same distance from Earth as the Cartwheel galaxy, likely passed through the latter earlier in its history. The collision disrupted the larger galaxy and sent shock waves outward, creating the circular structure. The waves also triggered the formation of the blue and white stars seen in the outer ring, as the waves compressed some pockets of gas and dust enough to trigger star formation.

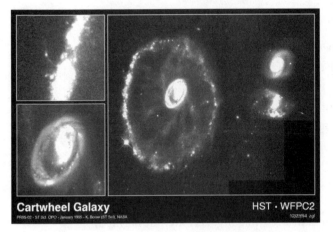

FIGURE 12.1
A Hubble Space Telescope image of the Cartwheel Galaxy, which appears to have been disrupted by the passage of the irregular galaxy in the lower right

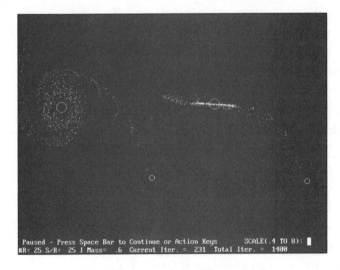

FIGURE 12.2
A screen shot from the program GalaxCrash Plus that shows two galaxies about to collide

While advanced modeling of galaxy collisions requires sophisticated models that run on supercomputers, you can get a sense of what galaxy collisions are like with Zephyr Software's GalaxCrash Plus for DOS and Windows. This program allows you to create models of colliding galaxies and observe the results (see Figure 12.2). The models aren't as sophisticated as those theoretical astrophysicists use on supercomputers, but the program still gives you

FIGURE 12.3
An image of NGC 1738 that shows a pair of spiral galaxies colliding

an idea how galaxies interact. More information about the software is available at **http://www.zephyrs.com/galax.htm**.

For images, perhaps the best resource is from the University of Alabama's astronomy department at **http://crux.astr.ua.edu/pairs2.html**. This site has a collection of images of colliding galaxies and information about the galaxies and the processes taking place—for example, NGC 1738, which is a pair of galaxies interacting (see Figure 12.3).

Active Galaxies

Active galaxies are unusually bright, or active, for their size. This brightness is usually constrained to the galaxy's core, which indicates that whatever process is making the galaxy so bright (an active galaxy can be 10 to 100 times as bright as our own Milky Way) is located there. What that process is, though, has been a major question in astronomy for decades.

The first active galaxies discovered were a set of spiral galaxies studied by Carl Seyfert in the 1940s. These spiral galaxies have unusually bright cores; a short-exposure image of such a galaxy shows a starlike image; to see the spiral arms one has to expose the image long enough to overexpose the core. The brightness of these objects can change over a few weeks or months by as much as the combined brightness of the Milky Way, which implies that whatever is making the galaxy incredibly bright must also be incredibly small: no more than a light-year in diameter. A good Web site with more information about Seyfert galaxies is at the University of Oregon, at **http://zebu.uoregon.edu/~soper/ActiveGalxy/seyfert.html**. More advanced, detailed information is available as part of a research report from Johns Hopkins University at **http://praxis.pha. jhu.edu/papers/papers/afdscirev_b/node9.html**.

Seyfert galaxies have recently been linked to quasars (see the next section for more information about quasars), so the best place to find images of Seyfert galaxies is in catalogs of quasar images. The Astronomy Picture of the Day archive at **http://antwrp.gsfc. nasa.gov/apod/lib/quasars.html**, includes a particularly good image

FIGURE 12.4
An image of NGC 3393, a Seyfert galaxy, from the Astronomy Picture of the Day archives; note the very bright core in the shape of an S

of a Seyfert galaxy, NGC 3393 (see Figure 12.4). The image shows a very bright center with bright arms trailing away in an S shape.

Another class of active galaxies is the *BL Lacertae* (or BL Lac) group, named after the first object of this type discovered. When BL Lac was first discovered in the late 1920s it was thought to be just a variable star, since it seemed pointlike in early images and its brightness varied by several magnitudes over just a few months' time, neither of which were known properties of galaxies at the time. Later discoveries, however, showed the object to be distinctly fuzzy, like a galaxy, and a spectral analysis of the light showed it to be like that of an elliptical galaxy; hence BL Lac's reclassification as a galaxy, along with dozens of other similar objects. An important difference between BL Lac objects and Seyfert galaxies is that the former are bright elliptical galaxies without any spectral lines in their light, while the latter are spiral galaxies that show strong emission lines from elements like iron.

Radio astronomy has played an important role in the identification of active galaxies. In the 1960s astronomers started finding unusual, powerful two-lobed sources of radio emissions in the sky. These lobes extended in a line on either side of a quiet central region. Radio telescopes could not detect anything between these lobes, but when placed on a map of the sky they appeared on either side of active galaxies, usually giant elliptical ones. The program Galaxy Guide for Windows shows a good example of these radio lobes from the radio source Centaurus A, superimposed on a large galaxy that appears to be the source of Centaurus A's radio lobes (see Figure 12.5).

These radio lobes form when electrons and magnetic fields ejected from the nucleus of the galaxy at a large fraction of the speed of light finally slow down, generating *synchrotron energy* (the radio energy formed when electrons travel through a spiraling magnetic field). To generate energy in this way requires a powerful source of energy in the galaxy's center.

More information about active galaxies in general is at **http://www.atnf.csiro.au/~bkoribal/definitions.html**, which provides

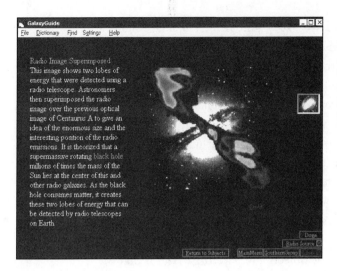

Radio Image Superimposed
This image shows two lobes of energy that were detected using a radio telescope. Astronomers then superimposed the radio image over the previous optical image of Centaurus A to give an idea of the enormous size and the interesting position of the radio emissions. It is theorized that a supermassive rotating black hole millions of times the mass of the Sun lies at the center of this and other radio galaxies. As the black hole consumes matter, it creates these two lobes of energy that can be detected by radio telescopes on Earth.

FIGURE 12.5
An image of the radio lobes of Centaurus A superimposed on the image of the galaxy that is its source, from the program Galaxy Guide for Windows

general descriptions of the different types, including some advanced ones not discussed here.

Quasars

Radio astronomy played a role in the discovery of active galaxies, and it has also helped uncover something even more unusual. The first surveys of the sky at radio wavelengths started in the mid-1930s by astronomer Grote Reber, who used a homemade telescope in his Illinois backyard. Early surveys found a number of strong radio sources; some of which were traced to supernovae, others to the nucleus of our galaxy, but one—dubbed Cygnus A, as it was the first radio source discovered in the constellation of Cygnus—could not be linked to any known object when it was first discovered. Only in the 1950s was a galaxy found in the same area of the sky as the radio source. The galaxy turned out to be very distant—some 750 million light-years away—which meant that the radio source had to be very powerful.

By the early 1960s many more radio sources had been discovered, and a few had been linked to objects at visible wavelengths.

These objects appeared starlike, and thus had to be close to us, perhaps within our galaxy. Their spectra, though, showed emission lines—wavelengths where the object was emitting light—that could not be identified with a particular process or element. In 1963, astronomer Maarten Schmidt of Caltech made the key breakthrough: The wavelengths were actually common hydrogen emission lines, normally found at much shorter wavelengths than were recorded for these objects. These wavelengths had been redshifted by the Doppler effect, the same effect that changes the pitch of a car horn from high (short wavelengths) as it approaches you to low (long wavelengths) as it moves away.

The redshift meant these objects were moving away; the large amount of redshift meant they were moving fast, and were thus far away. They had to be several billion light-years from Earth, making them some of the most distant objects ever observed. Since the starlike objects were visible in telescopes, they had to be incredibly bright, as bright as up to 100 galaxies. These objects became known as quasi-stellar radio sources, or simply *quasars*. A similar class of very bright, very distant objects that do not emit radio waves are known as quasi-stellar objects (QSOs), although these objects are usually lumped together with quasars.

More information about quasars is available online in a set of frequently asked questions posed by middle school students and answered by a professional astronomer at Virginia Tech. Check **http://www.phys.vt.edu/~astrophy/faq/quasars.html** for these questions and answers, which provide a good introduction to the subject. The bare basics are available at **http://chemlab.pc. maricopa.edu/astro/quasar.html**, which combines a few figures and some text to provide an at-a-glance explanation. More-detailed information is available at **http://astro.ph.unimelb.edu.au/ ~bholman/qso/qso1.html**, a set of class notes with links to other information and images. A student project at Rice University, **http://www.owlnet.rice.edu/~budmil/**, discusses the link between quasars and galactic evolution.

FIGURE 12.6

A Hubble image of the quasar PKS 2349

FIGURE 12.7

The hundred-thousandth image taken by the HST features a quasar (the bright object in the left-center portion of the image) as well as a pair of colliding galaxies, visible near the top

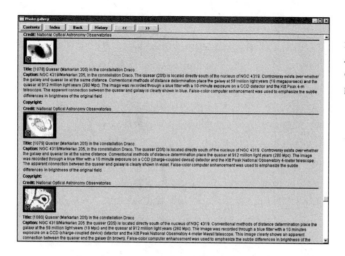

FIGURE 12.8
A portion of the gallery of quasar photos from the program Redshift

There are also several good collections of quasar images online. In addition to the Astronomy Picture of the Day archive mentioned earlier, at **http://antwrp.gsfc.nasa.gov/apod/lib/quasars.html**, the Hubble Space Telescope has an extensive archive of quasar images, listed at **http://oposite.stsci.edu/pubinfo/Subject.html#Quasars**. A good example is the Hubble image of quasar PKS 2349, available at **http://oposite.stsci.edu/pubinfo/PR/95/04.html** (see Figure 12.6). The hundred-thousandth image taken by the Hubble Space Telescope also featured a distant, bright quasar; it's at **http://oposite.stsci.edu/pubinfo/PR/96/25.html** (see Figure 12.7). This image has the added bonus of showing a pair of colliding spiral galaxies.

The program Redshift for Windows and Macintosh features an extensive gallery of quasar images (see Figure 12.8), including explanations of the objects and the processes that power them. The program also includes a detailed dictionary entry on quasars, with links to related entries in the program.

What powers these incredibly bright objects? The leading candidate for a quasar power source is a supermassive black hole in the center of a galaxy (see Chapter 9). Such a giant black hole would attract a massive accretion disk of dust and gas that would orbit the black hole and slowly spiral in towards it. As the material spirals in, its gravitational energy is converted into heat, which is in turn radiated as light. However, because of pressures within the accretion disk or magnetic fields, much of the material would not fall into the black hole itself but would be ejected at high speeds from a pair of jets on either side of the black hole. A *Scientific American* article, available online at **http://www.sciam.com/0297issue/ 0297scicit4.html**, discusses this link between quasars and supermassive black holes in greater detail. This model may describe not only quasars but also Seyfert galaxies, BL Lac objects, and radio galaxies. The more massive the black hole, the larger and more energetic its accretion disk will be. Thus a supermassive black hole that weighs 1 to 10 million times as much as our Sun could power a Seyfert galaxy, while a more massive black hole with up to a billion times the Sun's mass might be needed for a quasar. The viewing angle is also important. If we looked at the supermassive black hole right down one of its jets, we would see a fuzzy object with little sign of emission or absorption from the accretion disk: a BL Lac object. Tilt the viewing angle partway to the side, and we would see the bright area of the accretion disk and perhaps evidence of a jet coming out of the limb: a Seyfert galaxy or a quasar. Tilt our viewing angle to the side, and we would not see the accretion disk much at all, but we would easily detect the jets spewing forth from either side of the disk: the double lobes of a radio galaxy.

This explanation requires that quasars be linked to the cores of galaxies, something astronomers had assumed for many years. However, Hubble Space Telescope observations released in 1995 appeared to show several quasars not hosted in a galaxy; they would not have access to the engine of a supermassive black hole to generate their energy. These quasars do appear to be near other galaxies, but not close enough to be generated by these galaxies. No one has discovered any other powerful source of energy for a quasar. As John Bacall of Princeton University, who has studied quasars for more than 30 years, said, "This is a giant leap backward in our understanding of quasars." It's also an opportunity for new discoveries that may shape our understanding of the universe. These objects show that the universe can be a very interesting, even bizarre place.

But how did all of this—these planets, stars, galaxies, clusters, quasars, and more—come into existence in the first place? In the next chapter we'll take a look at the question of the origin of the universe, probably the biggest question in the history of astronomy.

13

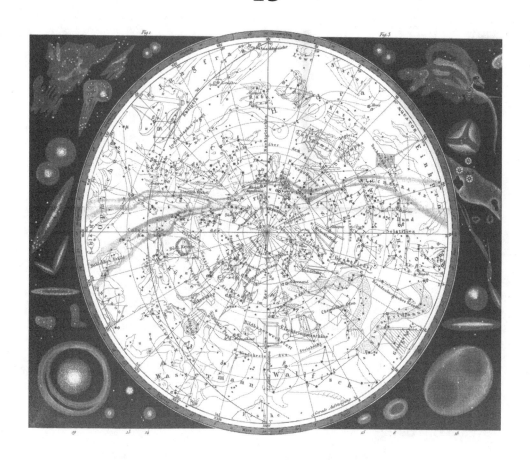

COSMOLOGY:
BEGINNINGS AND
ENDINGS

We see throughout the universe evidence of birth and death. Stars and planets are created out of clouds of gas and dust. Stars die, and in the process create new nebulae, neutron stars, and even black holes. Galaxies collide, destroying old galaxies and creating new regions of star formation in the remnants. The universe is full of beginnings and endings.

However, what about the universe itself? Was it born, or has it existed forever? Will it continue to exist for eternity, or will there be a day when the universe comes to an end? These questions, some of the most important in astronomy, and arguably some of the most important of human civilization, are studied in a branch of astronomy called cosmology. Cosmologists seek to understand whether and how the universe was created and evolved into its present state, and to determine its ultimate fate. In this chapter we'll look at some of the cosmology resources available in software and online.

The Big Bang

Albert Einstein laid the seeds for modern cosmology. In 1916 he unveiled his general theory of relativity, which explained gravity as a curvature of space-time, a combination of the three dimensions of physical space plus time. Einstein's relativity equations had one amazing aspect: they appeared to require the universe to be expanding or contracting. This so unsettled Einstein—who, like most scientists of the day, thought that the universe was static and unchanging—that he added an additional factor dubbed the *cosmological constant* to his equations to keep the universe static.

Observations soon caught up with theories, though. Edwin Hubble, who proved that many of the nebulae seen in the night sky are distant galaxies not dissimilar to our own, studied the spectra of these galaxies. He noticed that the bands in each spectrum were shifted, usually toward the red. By comparing these redshifts with the distances to the galaxies, he came across a surprising conclusion: the farther away a galaxy was, the greater its redshift. The Doppler effect caused these redshifts (see Chapter 12); the farther a galaxy was from Earth, the faster it was moving away. This relationship between galaxy distance and redshift is often called Hubble's law, and the ratio of the redshift velocity to the distance is called Hubble's constant.

The fact that galaxies are receding from Earth implies that at one time they were closer. At some point in the past, then, all the galaxies had to be close together. A simple way to find out how long ago that was is to divide 1 by the Hubble constant: using modern values, it works out to about 12 to 15 billion years. That raises the question: what was the universe like when everything was packed together?

One theory proposed in the 1940s, which gained wide acceptance among astronomers in following years, was that the universe was created billions of years ago in an explosion of scarcely imaginable proportions. The explosion created the universe and all the matter and energy within it; nothing existed before the explosion. Some astronomers, such as eminent British astronomer Sir Fred Hoyle, all but laughed at the concept; Hoyle dubbed this universe-creating explosion the Big Bang. The name stuck.

According to the Big Bang theory of the universe's formation, the universe quickly expanded and cooled after this initial explosion. As it cooled, matter began to form, first as quarks, the most basic building blocks of matter, and then more familiar subatomic particles such as protons, neutrons, and electrons. These combined to create atoms of hydrogen and helium, the two lightest elements. Much later, clouds of hydrogen and helium coalesced and started to form the first stars and galaxies. The universe continued to expand from the force of the Big Bang, explaining the receding galaxies that Hubble and other astronomers observed.

There are a great number of online sources of information about the Big Bang and cosmology in general. Cambridge University has a introduction to the Big Bang and a brief history of the universe at **http://www.damtp.cam.ac.uk/user/gr/public/bb_history.html**. Astronomer Dr. Sten Odenwald has a collection of essays about the Big Bang and cosmology that he wrote for major magazines at **http://www2.ari.net/home/odenwald/cosmol.html**. NASA's Goddard Space Flight Center provides a good, brief introduction to the Big Bang at **http://map.gsfc.nasa.gov/html/big_bang.html**. The University of Illinois's Cosmos on a Computer Map, at **http://141.142.3.76/Cyberia/Expo/cosmos_nav.html**, is a detailed introduction to cosmology in general and the Big Bang in particular, complete with graphics and animations to illustrate key points.

A good way to learn about the Big Bang and cosmology is to go over class notes from courses on the subject. A set of notes on cosmology by astronomy professor Nick Strobel is available at **http://www.bc.cc.ca.us/programs/sea/astronomy/cosmolgy/cosmolgy.htm**. A tutorial on cosmology by UCLA astronomer Ned Wright is at **http://www.astro.ucla.edu/~wright/cosmoall.htm**. You can view slides from a lecture on the Big Bang from a course at the University of Rochester at **http://www.pas.rochester.edu/~dmw/Lect_21/sld001.htm**.

In addition to these basic resources about the Big Bang and cosmology, more advanced, detailed information about the subject is available online. A report by the National Research Council, "Cosmology: A Research Briefing," goes into great detail on cosmology. The online version, available at **http://www.nap.edu/readingroom/books/cosmology/**, includes the full text as well as figures and images used in the original printed report. If you're looking for sources of information about the Big Bang offline, first check a reading list on the subject at **http://physics7.berkeley.edu/darkmat/reading.html**.

Cosmology does not produce the pretty pictures that other fields of astronomy create, but there are still some good images related to the field. The Hubble Space Telescope maintains a list of cosmology-related images on its Web site at **http://oposite. stsci.edu/pubinfo/Subject.html#Cosmology**. Among them is a new image of the most distant galaxy ever seen, a dim red galaxy that may have formed when the universe was just a billion years old. The image, and information about it, is available online at **http://oposite.stsci.edu/pubinfo/PR/97/25.html** (see Figure 13.1).

Computer programs provide an additional source of information about the Big Bang. Discover Astronomy, by Maris Multimedia for Windows, features an animated guided tour titled Big Bang to Galaxies that shows the evolution of the universe from the Big Bang to the formation of modern galaxies (see Figure 13.2). Redshift 2, also by Maris Multimedia for Macintosh and Windows, features the same guided tour as well as a number of dictionary entries that provide a basic introduction to cosmology and the Big Bang. Similar information is available from Galaxy Guide, by Parallax for Windows.

Perhaps the most comprehensive software package about the Big Bang is The Universe: From Quarks to the Cosmos, by Sci-

FIGURE 13.1
The most distant galaxy known to exist in the universe appears in this Hubble Space Telescope image

entific American Library for Macintosh and Windows. The program features a detailed essay on the Big Bang and shows models that let you see the formation and evolution of the universe since the Big Bang (see Figure 13.3). The program also features interviews with astronomers and physicists about Big Bang–related topics as well as Themes, narrated tutorials on a number of topics related to the Big Bang and cosmology.

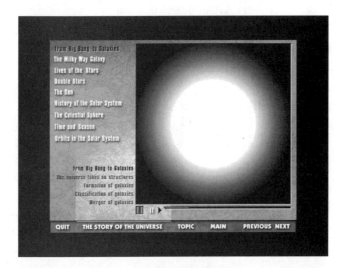

FIGURE 13.2
Discovery Astronomy provides this tutorial about the Big Bang and the early universe

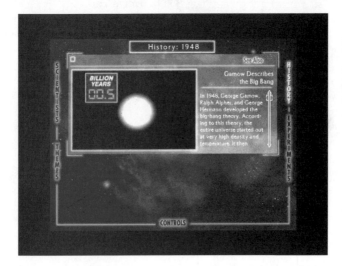

FIGURE 13.3
The Universe shows how the universe evolved from the Big Bang to the present day distribution of galaxies and stars

Evidence for the Big Bang

While the Big Bang theory does explain the expansion of the universe that relativity predicted and astronomers saw, that alone is not sufficient to prove the theory. Considerable work has been invested into research to prove–or disprove–the Big Bang concept. Much of this work has focused on one particular prediction of the Big Bang theory: that an initial burst of radiation from the Big Bang, at the time intensely hot, should have expanded with the universe and cooled, creating a faint signature observable everywhere.

In the early 1960s a group of cosmologists at Princeton University began to construct a search for this cosmic background radiation, which they believed would be visible at microwave wavelengths of light, much longer than what is visible by humans, corresponding to a temperature of just a few degrees above absolute zero.

While they planned their search, though, a pair of Bell Labs physicists, working just a few miles away from Princeton, were puzzled by the behavior of a radio antenna designed to relay telephone calls to new orbiting communications satellites. Wherever they pointed their telescope in the sky, they detected an unusual background signal. Thinking the problem was with their antenna, they performed exhaustive tests of their equipment, and even scrubbed bird droppings from the interior of the horn-shaped antenna. They could find nothing wrong, and yet the strange background noise persisted.

Eventually the Bell Labs physicists, Arno Penzias and Robert Wilson, got into contact with the Princeton team, who told the two about their plans to look for the cosmic background radiation. After sharing information, though, both groups realized Penzias and Wilson had already discovered it. Other researchers soon confirmed their work, using other radio telescopes. A key signature of the Big Bang had been discovered, providing strong evidence that the Big Bang did take place. Penzias and Wilson won the Nobel Prize for their discovery.

Some basic information about the cosmic background radiation (also called the cosmic microwave background radiation) is available online. Introductions to cosmic background radiation are at **http://spectrum.lbl.gov/WWW/cmb.html** as well as **http://map.gsfc.nasa.gov/html/cbr.html**. The basics of cosmic background radiation, with links to other sources of information online, is at **http://cmbr.phys.cmu.edu/cmb.html**. Some class notes about cosmic background radiation are at the University of Oregon at **http://zebu.uoregon.edu/~soper/Early/expansion.html**. A list of frequently asked questions and answers about cosmic background radiation is available at **http://www.astro.ubc.ca/people/scott/faq_basic.html**.

While the existence of cosmic background radiation had been confirmed, astronomers were still interested in studying it. In particular, they wanted to detect any slight variations in the radiation that could be linked to clumps of matter and energy in the early universe, from which the first stars and galaxies could have spawned. If the background radiation was perfectly smooth, it would be difficult to explain how matter came together to form the universe we see today, and thus would be a setback for the Big Bang theory.

In 1989 NASA launched the Cosmic Background Explorer (COBE), a satellite whose purpose was to map the intensity of the cosmic background radiation to a level of sensitivity never before achieved. Its first results confirmed that the background radiation had the same temperature throughout the sky, at the predicted level of about 2.74 degrees Kelvin (about -454° Fahrenheit). However, after carefully adding up millions of observations from COBE, scientists found ripples in the background radiation: differences in temperature of only 0.01 percent. The effect is small, but appears to be significant enough to show that there were differences in the early universe's density that could have helped establish the formation of stars and galaxies.

The official COBE Web site, at **http://www.gsfc.nasa.gov/astro/cobe/cobe_home.html**, has information about the spacecraft and its mission, including images of the spacecraft and its data (see Figure 13.4). Lawrence Berkeley Laboratory, home of some of the scientists who worked on COBE, has its own COBE Web site at **http://spectrum.lbl.gov/www/cobe/cobe.html**, which includes a temperature map of the background radiation, showing the differences in temperature across the sky (see Figure 13.5). Some general information about COBE is also available from NASA's Spacelink educational service at **http://spacelink.nasa.gov/NASA. Projects/Space.Science/Universe/Cosmic.Background.Explorer. COBE/index.html.**

The discovery of the cosmic background radiation was a major piece of evidence in support of the Big Bang, but it also caused problems for the theory. Portions of the universe were believed to be more than 20 billion light-years apart, based on the redshifts of

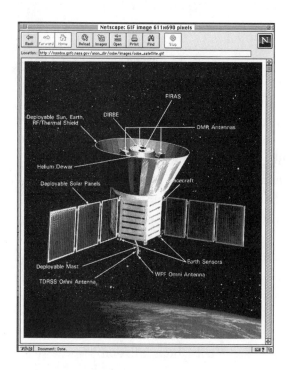

FIGURE 13.4
An illustration of the COBE spacecraft that provided the most detailed map of the cosmic background radiation to date, from the COBE Web site

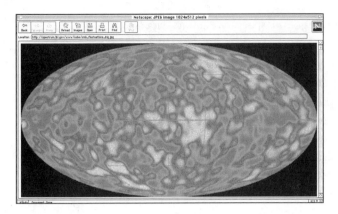

FIGURE 13.5
The map of the cosmic background radiation produced from COBE data, which shows the very small variations.

galaxies going in opposite directions from the Earth, yet these regions had the same cosmic background temperature. Since the universe was believed to be much less than 20 billion years old, these distant regions could not have been in contact with each other since the Big Bang, yet they had essentially the same temperature. How could this be?

A solution to this problem came in the early 1980s from researcher Alan Guth. He proposed that the universe underwent a period of inflation very early in its history. This inflation caused the size of the universe to expand by an extraordinary amount—up to a factor of 10^{50}, a 1 followed by 50 zeros—in a tiny fraction of a second, less than a second after the Big Bang. This pushed different parts of the universe far apart, too far for them to see each other. Thus, areas that were once close together and had the same temperature were now far apart, creating the background radiation pattern we see today.

There is a lot of information about inflation available on the Web. Introductory information about the horizon problem mentioned above, and its solution through inflation, is available at a couple of sites, **http://spectrum.lbl.gov/WWW/horizon.html** and **http://map.gsfc.nasa.gov/html/inflation.html**.

Cambridge University provides more details about inflation and its relevance to cosmology at **http://www.damtp.cam.ac.uk/user/gr/public/inf_prob.html**. A more advanced explanation of inflation for those familiar with research in the field and not afraid of mathematics is at **http://galaxy.cau.edu/tsmith/cosm.html**. British science writer John Gribbin has written perhaps the most detailed yet readable account of inflation and its connection to our current understanding of cosmology. His account of inflation is on the Web at **http://epunix.biols.susx.ac.uk/Home/John_Gribbin/ cosmo.htm**.

Alternative Theories

While the Big Bang has become the leading and most widely accepted theory about the universe's formation, one that explains much of modern cosmology, it is not universally accepted. Others continue to believe that the Big Bang does not adequately describe the beginning of the universe, and seek other explanations.

One astronomer who has criticized the Big Bang theory for nearly 50 years is British astronomer Sir Fred Hoyle. Hoyle, who derisively gave the Big Bang its name, which has gained common acceptance, proposed in the late 1940s an alternative, the steady-state theory. Hoyle and his colleagues tried to explain the expansion of the universe by the creation of new matter. Thus the universe continues to expand, pushed out by the new matter, without the need for an explosion like the Big Bang. However, neither Hoyle nor anyone else could explain how the matter was created, and as evidence such as the cosmic background radiation was discovered that supported the Big Bang, but not the steady-state theory, support for Hoyle's explanation waned.

Still, there are those today who do not accept the Big Bang theory and seek alternate explanations. One explanation debunks the use of redshifts as accurate measures of distance by claiming that light becomes "tired" and turns redder over distance, thus exaggerating the distances that redshifts measure. Another explanation tries to make the formation of plasmas in the early universe prominent in the creation of large-scale structures, thus making inflation and temperature variations in the background radiation unnecessary.

Some of the problems with and alternatives to the Big Bang theory are available online as part of the notes for a cosmology class at **http://www.kingsu.ab.ca/~brian/astro/chp20.htm**. Of course, the Big Bang theory has plenty of devoted supporters who provide their own arguments supporting the Big Bang. UCLA's Ned Wright provides one detailed list of supposed errors in attacks on the Big Bang at **http://www.astro.ucla.edu/~wright/errors.html**.

The Fate of the Universe

Even if the astronomical community completely accepted the Big Bang, it leaves open one critical question: What is the eventual fate of the universe? Will it continue to grow and expand, or will it one day come to a stop, and perhaps start to shrink, leading to what some have called the Big Crunch? This is one of the key questions in cosmology today, and has not yet been satisfactorily answered.

The fate of the universe depends on the density of matter in the universe, which in turn determines how much of an effect the force of gravity will have on the universe's expansion. If the density is too low, gravity will not be able to stop expansion, and the universe will continue to expand forever, growing larger and colder: an open universe. If the density is too high, the force of gravity between objects in the universe will bring the expansion to a stop, and then the universe will begin to contract, eventually leading up to the Big Crunch—a form of the Big Bang in reverse, a closed universe. If the density is just right, at a single critical value, the universe's expansion will come to a stop, but will take an infinite amount of time to do so: a flat universe. Such a universe would last forever, but would slowly stop expanding. The program The Universe, from Scientific American Library for Macintosh and Windows, features an animation that explains what each of the three possibilities would look like (see Figure 13.6). The Space Telescope Science Institute, operators of the Hubble Space Telescope, also have a good illustration showing the different possibilities available from **http://oposite.stsci.edu/pubinfo/PR/96/21.html** (see Figure 13.7).

The Big Bang theories that rely on inflation, which is commonly accepted today by astronomers, require the universe to have that "just right" density of matter to work. But the actual density of matter in the universe today is thought to be much lower, implying the universe would instead expand forever. Astronomers are searching

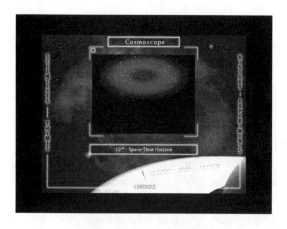

FIGURE 13.6
A scene from the program The Universe, from Scientific American Library, which shows the Big Crunch, one of three possible fates for the universe

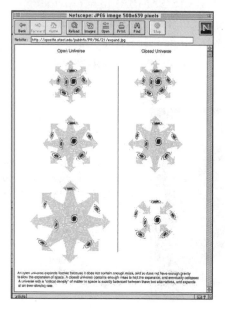

FIGURE 13.7
An illustration showing the universe's possible fates, from the Space Telescope Science Institute Web site

for the missing matter in the universe that would increase its density to fit the Big Bang models. However, new data suggest that the expansion of the universe is not slowing down, as once thought, but accelerating. Two teams of astronomers have found that supernovae in distant galaxies are much farther away from us than thought from their redshifts (Figure 13.8). They interpret this to believe that something is causing the universe to expand faster now than it did in the early universe. You can check their results online at **http://www-supernova.lbl.gov** and **http://cfa-www.harvard. edu/cfa/oir/Research/supernova/HighZ.html.**

Future spacecraft missions may help determine whether the universe is open, closed, or flat. The Microwave Anisotropy Probe, or MAP, a planned NASA mission, is a more advanced version of the COBE mission, with more sensitive detectors. It will be better able to observe differences in the cosmic background radiation and make other observations that would show what type of universe we live in. Information about MAP, scheduled for launch in the fall of the year 2000, is at **http://map.gsfc.nasa.gov/.** A similar European mission called Planck (formerly COBRAS/SAMBA) may provide even more accurate measurements, but would not launch until about 2005, assuming the mission is funded. Information about Planck is at **http://astro.estec.esa.nl/SA-general/Projects/Planck.**

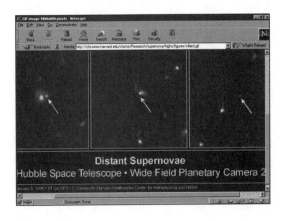

FIGURE 13.8
Observations of distant supernovae like these seen by the Hubble Space Telescope are raising new questions about the fate of the universe

It may be years before we have a good idea of how the universe will end, or even a better understanding of how it was created. However, these questions are among the most significant in all astronomy, and for all humanity, for they give us a clear view of how the very time and space in which we exist came into being, and how it might end.

With our virtual tour of the universe from Earth to the Big Bang completed, we now turn our attention to examining in detail some of the computer tools you can use to learn more about these subjects. The next chapter will look at the various software packages available for computers on all aspects of astronomy, while the chapter after that will look in detail at the online resources, both on the Web and online services, that provide information about astronomy.

14

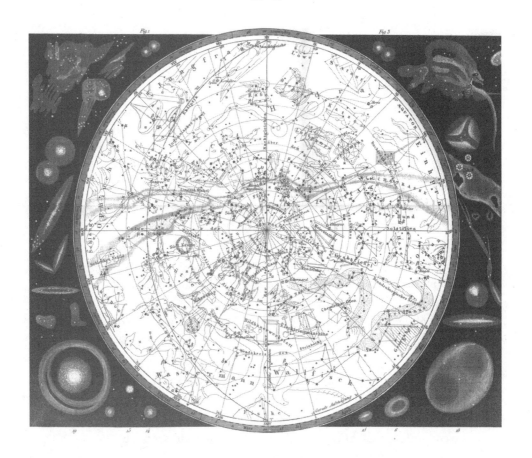

EXPLORATIONS ON YOUR OWN: OFFLINE

There is no shortage of software available for personal computers that enables you to learn more about astronomy. There are dozens of programs that will show you the sky at any given time and location on the planet, giving you the ability to find the location of a particular star or planet in the sky. There are many other programs designed to teach users about some specific aspect of astronomy, from the Solar System to the Big Bang.

Astronomy software can fill a variety of roles for different audiences. For the amateur astronomer, for example, software offers a way to pursue your hobby even when the sky is overcast. Dreary nights might be ideal for doing historical astronomical research, planning observation or astrophotography sessions, running "what if" simulations, viewing the skies of past and future times, or simulating events such as the impact of comet Shoemaker-Levy 9 on Jupiter. On clear nights, astronomical software might prove useful for guiding a computerized telescope to a specific faint and obscure deep-space object, printing star charts for a specific observation session, controlling a CCD camera, or simply seeing what is available for viewing with binoculars or the naked eye.

This chapter provides some information about many of the astronomy programs currently available for Macintosh and Windows computers. The list here is not meant to be comprehensive; there are other programs out there, and new programs appear regularly, but what's listed in this chapter is certainly a very large sample of what's available, and it includes the best programs currently available. A larger list of astronomy software, including some older programs that may no longer be available, is maintained by *Sky and Telescope* contributing editor John Mosley and is online at **http://www.skypub.com/resources/software/commercial.html**.

The CD-ROM that accompanies this book includes some samples of the software listed in this chapter, either as shareware programs, which you can try out and, if you're satisfied, purchase from their creators; or demo versions, which show some of the features of the full-fledged versions. In addition, the CD-ROM has a vast collection of other shareware and freeware astronomy programs, some of which serve specialized needs or work well on older computers.

Planetarium Software

A large fraction of astronomy software can be classified as planetarium programs. As the name suggests, they display views of the night sky, which you can alter by changing the time and location. Many programs, though, go far beyond simply displaying images of the sky. They can locate and provide information on thousands or even millions of stars and other objects in the sky, and can simulate views the ordinary observer may never see, such as looking at Earth from space. Some programs provide detailed information on specific classes of objects, such as asteroids and deep-sky objects, while others can interface with specific types of telescopes and control them, allowing an observer to plan an evening's observations with just a few mouse clicks.

The sophistication (and price) of planetarium software ranges greatly. A number of planetarium programs are described here to help you decide which program best meets your observational needs.

Tours of Some Planetarium Packages

A number of excellent programs serve the needs of beginners while providing more advanced users with the information they need. **Redshift 3** (Piranha Interactive, 602/491-0500, **sales@ piranhainteractive.com**, **http://www.redshift3.com/**) includes several guided tours and tutorials that teach beginners the basics of astronomy from the Solar System to the Big Bang, as well as how to find objects in the night sky. More advanced users will benefit from a database of a quarter million stars and 40,000 deep-sky objects

available on the CD-ROM (see Figure 14.1). The program also features the full text of Penguin's *Dictionary of Astronomy,* complete
with hyperlinked cross-references and hundreds of images of planets, stars, and galaxies. Redshift 2 has a number of gee-whiz
features, such as the ability to view the sky from locations other
than Earth's surface: for instance, from other planets or deep space.
The program is available for Windows and Macintosh machines. A
similar program is **Distant Suns** (Monkey Byte Development,
800/522-3774, **http://www.distantsuns.com**).

Advanced users will benefit from information on over 16 million
stars from the Hubble Space Telescope (HST) Guide Star Catalog
accessible in the program. It includes hundreds of images and video
clips. The program can also remind users of upcoming astronomical events, such as eclipses, and keeps a monthly calendar of Moon
phases and other astronomical phenomena. Distant Suns is available for Macintosh and Windows computers and is bundled with
the Mars Rover program (see the Mars Rover listing that follows).

Macintosh users may enjoy **Voyager II** (Carina Software,
510/355-1266, **http://www.carinasoft.com**). The program includes
tens of thousands of stars and deep-sky objects and comes with a

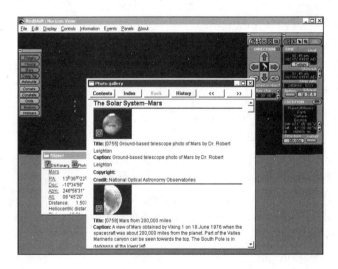

FIGURE 14.1
Redshift displays infor-
mation on Mars while
showing the night sky

CD-ROM with the millions of stars of the HST Guide Star Catalog for additional detail (see Figure 14.2). The program includes the orbits of well-known asteroids and comets and allows users to add the orbits of new objects. Version 2.0 of Voyager II is available for the Macintosh; Voyager III should be released by the time you read this, with both Macintosh and Windows versions.

You'll find it hard to beat the excellent graphics and features found in **Starry Night Deluxe** (Sienna Software, 416/926-2174, **contact@siennasoft.com**, **http://www.siennasoft.com**) for Macintosh and Windows systems (see Figure 14.3). The program includes stars from the HST Guide Star Catalog and thousands of deep-sky objects, with images of many. There are detailed images and illustrations of planets; you can zoom in and see them in close detail, as well as adjust your viewing location from locations on Earth and the Solar System. You also have access to fun bells and whistles, such as the ability to personalize the scenery around your observing site. The program comes with the LiveSky plug-in to allow you to control specific telescopes, such as the Meade LX-200, from the program. In addition to Starry Night Deluxe, a less expensive Basic

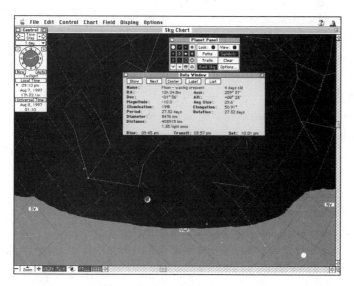

FIGURE 14.2
Studying the evening sky using Voyager II version 2.0

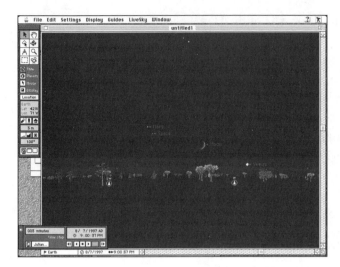

FIGURE 14.3
A view of an early evening sky, with the Moon, planets, and stars visible, from Starry Night Deluxe

version, with a subset of some of the features described here, is available.

Windows users can try **Expert Astronomer** (Expert Software, 305/567-9990, **sales@expertsoftware.com**, **http://www. expertsoftware. com**) for an inexpensive introduction to astronomy. The planetarium program features over 10,000 stars and other objects. However, its interface and controls are not as good as those of many of the other programs already described. It also features a large collection of images and video clips. The program works in Windows 3.1 and 95.

SkyChart III (Southern Stars Software, 408/973-1016, **info@ southernstars.com**, **http://www.southernstars.com**) is another excellent and reasonably priced application that includes a wealth of features. In addition to providing an excellent planetarium function (Figure 14.4), SkyChart III can compute and display the positions of Earth satellites using standard NASA/NORAD two-line element satellite orbit files and print high-resolution star charts and highly detailed ephemerides. The application also includes a fully customizable object database containing more than 40,000 stars and hundreds of Messier, NGC, and IC deep-sky objects. All existing objects can be edited, including the constellations. You can import your own database to add new comets, asteroids, stars, planets, or deep-sky objects. Especially nice features of SkyChart III are its ability to accurately reproduce eclipses, transits, and occultations—as seen not only from Earth, but from any place in the Solar System—and to show the sky dynamically from any object in the Solar System and beyond. This dynamic functionality can also let you watch the celestial poles precess and the constellations shift as the stars move over the course of thousands of years.

There are also good shareware programs that provide views of the night sky. **MPj Astro** (Microprojects, 416/221-8579, **mpj@total.net, http://www.total.net/~mpj/pages/mpj.html**) includes stars down to the sixth magnitude—the limit of naked-eye observing—as well as the planets and some of their moons. Those who register get access to an additional database of stars that goes down to the eighth magnitude and a collection of high-resolution images. MPj Astro is available for the Macintosh for a registration fee.

Deepsky 99 (Steven S. Tuma, 608/752-8366, **stuma@inwave.com, http://www.deepsky2000.com**) could more accurately be described as a collection of integrated applications, one of which happens to be a planetarium program (Figure 14.5). Here you will find image processing tools, customizable databases, an observation

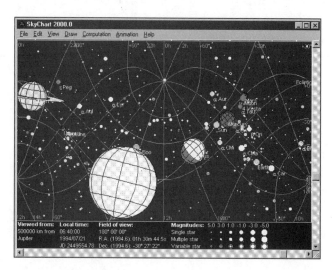

FIGURE 14.4
A map of how the sky might look a half million kilometers from Jupiter, as seen in SkyChart III

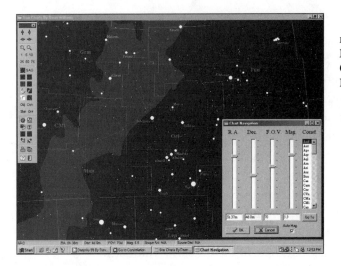

FIGURE 14.5
Exploring the constellation
Orion and nearby stars in
Deepsky99

session planner, the ability to drive computer-controlled telescopes, and more. Deepsky 99 is a labor of love for its author Steven Tuma and is offered as shareware for Windows 95, 98, and NT.

CyberSky (Stephen Michael Schimpf, 310/530-6766, **CyberSky@ compuserve.com, http://cybersky.simplenet.com**) is a shareware planetarium program for Windows users (see Figure 14.6). It includes stars down to the sixth magnitude and the planets of the Solar System. It is available in versions for Windows 3.1 and 95 for a registration fee.

Advanced Observation Programs

The programs described in the preceding section will meet the needs of many, if not most, people, but some users are looking for additional features to plan observations. Several programs target the dedicated amateur astronomer, rather than the causal observer, providing additional features or focusing in depth on a specific type of object, such as asteroids or deep-sky objects.

The Sky (Software Bisque, 800/843-7599, **thomasb@bisque. com, http://www.bisque.com**) is an excellent tool for amateur astronomers. Although beginners and casual observers will get a lot

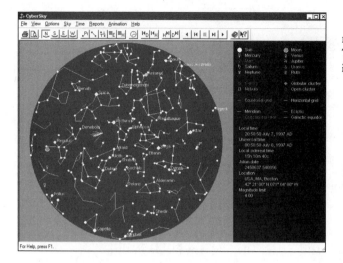

FIGURE 14.6
The night sky as seen
in CyberSky

out of this program, The Sky is truly designed for dedicated observers; it certainly falls into the "Swiss Army Knife" category and is perhaps the most feature-rich of all the readily available commercial astronomy applications. The program supports almost 20,000,000 stars from the HST Guide Star Catalog and has additional support for objects in the Digitized Sky Survey (available separately in 8 and 102 CD-ROM versions). The program can interface with specific telescopes for computer control of observations and can (with the CCDSoft program, available separately from the same company) process any images that an observer might take with a CCD camera attached to the telescope (Figure 14.7). The focus on observers is evident in even little details, like the ability to display the screen in shades of red only (to preserve night vision).

In addition to driving computerized telescopes both locally and across the Internet, version 5.0 can also track artificial Earth satellites using standard two-line elements that are readily available from several online sources. Unfortunately, a demo version of The Sky is not available, but the Software Bisque Web site has a wealth of information on the company's applications and the equipment requirements necessary to run them. Updates, additions, and fixes

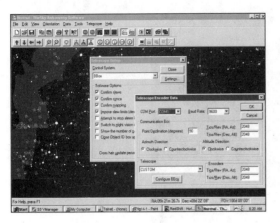

FIGURE 14.7
Setting up controls for operating
a telescope using The Sky

are posted fairly regularly at the site. The Sky is available in three
versions, Levels II, III, and IV, with an increasing number of stars
and deep-sky objects in each version. The software runs on Windows 95 and NT, and the company recently announced and
shipped a Macintosh version.

Those who are specifically interested in observing deep-
sky objects such as galaxies and nebulae will want to check out
NGCView (Rainman Software, 888/261-1686, **ngcview@rainman-
soft.com, http://www.rainman-soft.com**). The program includes a
database of 13,226 deep-sky objects, with detailed information
about each. Observers can use the program to check when and
where an object will be visible in the night sky (Figure 14.8). You
can use NGCView to generate a list of objects to observe on a given
night and then log your observations into the program for future
reference. NGCView is available in Windows 3.1 and 95 versions.

Those whose interest in deep-sky objects is limited to just the
100-plus objects in the Messier catalog may be interested in the
L.F.K. Messier Observer's Guide (Larry F. Kalinowski, 810/776-
9720). This program provides basic information on each Messier
object, including how it would appear through a small telescope.
This shareware program is written for DOS, but will run under
Windows 3.1 and 95.

The Messier Logging System (MLS) (Luis Argüelles, **whuyss@
tripod.net, http://members.tripod.com/~whuyss/mls.htm**) is a

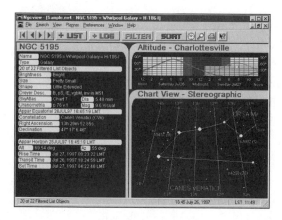

FIGURE 14.8
A screen of information about
NGC 5195, including a star chart,
observability information, and
more, from NGCView

free, customizable database of the Messier objects, written by Luis
Argüelles and made available on the Internet. For each entry, the
MLS provides the Messier number, the constellation where the
object is located, the celestial coordinates in right ascension and
declination, a short description of the object, its magnitude, and
various additional information. For identification purposes, an
image of each Messier object is included. The MLS is especially
useful as an observational aid, since personalized observation information can be appended to each object listed. This makes the MLS
a valuable tool for those interested in observing, learning about,
and exploring the brighter deep-space objects found here.

Those who want to observe asteroids will find **MPO99** (Bdw
Publishing, 719/748-8309, **71511.515@compuserve.com**, **http://
ourworld.compuserve.com/homepages/brianw_mpo**) a very useful tool. The CD-ROM includes detailed information on over
7,000 asteroids, which can be sorted by a variety of categories.
Users can generate lists of asteroids to observe each night and view
star charts to see where the asteroids will be. The program is available for Windows 3.1 and 95 users.

Asteroid Pro (Pickering Anomalies, **sales@anomalies.com**,
http://www.anomalies.com) offers features similar to MPO99 and
detailed information on thousands of asteroids. Its particular
strength is the ability to search for occultations of stars by asteroids;

astronomers observe these events to determine the size of the asteroid. Several levels of Asteroid Pro are available, with the more expensive and complete versions including 6,600,000 stars from the Hubble Space Telescope Guide Star Catalog (to about magnitude 13.5) to aid in occultation studies.

The Multiyear Interactive Computer Almanac (MICA) (Willmann-Bell, Inc., 800/825-7827, **http://www.willbell.com**) is produced by the U.S. Naval Observatory and marketed by Willmann-Bell. MICA provides high-precision astronomical data in tabular form for a wide variety of objects over a 15-year period (1990 to 2005). MICA is available both as a DOS application (that will run under Windows) and as a fully graphical Macintosh application. The algorithms used by MICA calculate much of the information tabulated in the annual publication *The Astronomical Almanac*. Numerous types of positions for celestial objects; times of the rise, set, and transit of celestial objects; and times of civil, nautical, and astronomical twilight are computed. MICA can also provide data on the apparent size, illumination, and orientation of the Sun, Moon, and major planets.

Other Planetarium Software Packages

We've described just a handful of the planetarium programs available for Macintosh and Windows; here is a more complete list.

Dance of the Planets (ARC Science Simulations, 800/759-1642, arcmail@arcinc.com, **http://www.arcinc.com/dance.htm**), written for DOS but capable of running in Windows 3.1 and 95, offers detailed views of planets and their moons, comets, asteroids, and other Solar System phenomena. The program is available in two versions; the more expensive version has more features and more objects in its database. The program has not been updated since 1994, as the company has turned to other projects, but it's still available for sale.

Deep Space (David Chandler Co., 209/539-0900, **david@davidchandler.com**, **http://www.davidchandler.com**) provides an alternative for DOS users. In addition to its planetarium features, Deep Space includes an outstanding observation guide, generates publication-quality star maps, serves as an almanac and ephemeris, provides an observation log, controls telescopes, and more. The newest version of Deep Spaces includes a shaded representation of the Milky Way and a full-year Moon phase calendar and has the ability to project local horizons onto the map of the sky (Figure 14.9). A CD-ROM version includes the millions of stars in the Hubble Space Telescope Guide Star Catalog and additional deep sky, comet, and asteroid databases.

Earth Centered Universe (Nova Astronomics, 902/443-5989, ecu@fox.nsta.ca, http://www.nova-astro.com) is a shareware Windows program equally useful to the armchair astronomer and the observing amateur astronomer. The CD-ROM version of the application includes a database of over 15,000,000 stars, the planets, the Sun, the Moon, comets, over 30,000 asteroids, and more than 10,000 deep-sky objects (Figure 14.10). This application can print high-quality star charts using any Windows-compatible printer and also controls modern computerized telescopes such as the Meade LX-200 series. It supports computer control of some other telescopes as well.

FIGURE 14.9
A star map, with a local horizon line plotted, from Deep Space for DOS

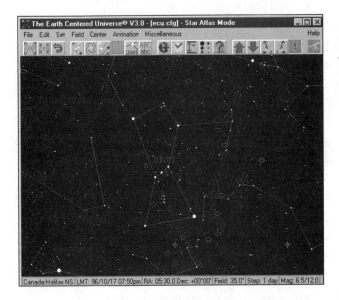

FIGURE 14.10

A view of Orion, including the many deep-sky objects visible in that region of the sky, from Earth Centered Universe

Guide 5.0 (Project Pluto, 800/777-5886, **pluto@projectpluto. com, http://www.projectpluto.com**) uses a database of over 15 million stars, thousands of deep-sky objects, and thousands of asteroids and comets. The program is available for PCs running DOS or Windows.

MegaStar (E.L.B. Software, 713/541-9723, **megastar@flash. net, http://www.flash.net/~megastar**) features the HST Guide Star Catalog and over 100,000 deep-sky objects on a CD-ROM. The program can also interface with some telescopes to provide computer-controlled observation sessions. It is available for PCs running Windows 3.1 or 95.

MacStronomy 2 (Etlon Software, 303/702-9274, **info@etlon. com, http://www.etlon.com/MacStronomy/Features.html**) is a Macintosh planetarium program that features the planets and all stars visible to the naked eye; additional catalogs (available separately) can expand the program to include hundreds of thousands of stars.

Observer (Procyon Systems, 407/366-5985, **observer@procyon-sys.com, http://www.procyon-sys.com**) is a Macintosh program that can generate star charts from a database of over 300,000

objects. The company also sells Scopelink, a hardware adapter that allows the program to control a variety of different telescopes. Observer is available in two versions; the more expensive version has additional capabilities.

PC-Sky (CapellaSoft, 888/422-7355, **crinklaw@n2.net**, **http://www.skyhound.com/cs.html**) is designed for amateur astronomers and provides detailed observation information on objects in the night sky. The program shows views of the sky as seen by the naked eye, binoculars, and telescopes. It is available for Windows 3.1 and 95 users. A more advanced program, **SkyTools**, is also available. It provides powerful searching and planning features for amateur astronomers.

Skychart 2000.0 (Southern Stars Software, 408/973-1016, **info@southernstars.com**, **http://www.southernstars.com/SkyChart.html**) is the latest version of the SkyChart program. It includes a database of thousands of stars and deep-sky objects as well as detailed data on Solar System objects. The program is available for Windows and Macintosh systems.

SkyMap Pro 5 (SkyMap Software, **sales@skymap.com**, **http://www.skymap.com**) includes over a quarter million stars and deep-sky objects, with detailed information useful for observers. The program is available for Windows systems.

Stargaze (CSB Metasystems, 800/232-7830, **cebs@star-gaze.com**, **http://www.star-gaze.com**) is a Windows program that features a database of a half-million stars and thousands of deep-sky objects. It is available for Windows computers.

Satellite Tracking Software

For many, the fascination with the night sky has to do not only with the relatively static stars and planets that appear to the naked eye, but with the moving points of light that are the artificial Earth satellites launched by man, ranging from the Space Shuttle and Mir space station to small, relatively unknown satellites. Such Earth satellite tracking software is dependent upon regularly updated positional data that is used to locate and track these artificial moons. Fortunately, there are several sites on the Internet that make such information available regularly and on a timely basis. The primary differences between the applications discussed in this section have to do with the computer system and operating system on which they run and the map projections and number of satellites they can track at a given time.

WinTrak Pro (Paul E. Traufler, 256/837-0084, **wintrak@traveller. com**, **http://www.hsv.tis.net/~wintrak**) tracks orbiting satellites as well as the Moon, Sun, and stars; it can continually update the screen with the current positions of multiple satellites and tracking stations (Figure 14.11). WinTrak Pro can be used to predict upcoming satellite passes, use real-time or simulated-time satellite tracking modes, provide constellation views for pass planning, and display fully rendered 3D Earth views in any video mode supported by Windows 95, 98, or NT. High-resolution maps can be printed to any printer supported by Windows. Some of the more esoteric functions of WinTrak Pro include Doppler shift and squint angle calculations and an interface to the Auto Tracker for automatic

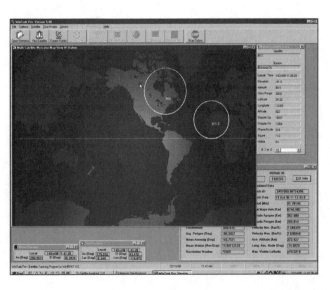

FIGURE 14.11
Tracking the Hubble Space Telescope (HST) and the Russian space station Mir in WinTrak Pro

antenna pointing and radio tuning. Also available from Paul Traufler are WinTrak for Windows 3.1 and TrakSat for DOS; the shareware version of TrakSat for DOS on the accompanying CD-ROM.

LogSat Professional (LogSat Software Corporation, 407/275-0780, **sales@logsat.com**, **http://www.logsat.com**) is a satellite tracking program for Windows 3.1, 95, 98, and NT that can track up to 2,000 satellites in one window, with several windows open at once, each independent of the others. Each window can be assigned a time and date specific to that particular window. Five different map projections are available in LogSat Professional: Mercator, Equidistant, Sinusoidal, Hammer, and Orthographic (Figure 14.12). The program provides an extensive set of tracking data, and single and multi-satellite schedules can be printed. The program can also display 3D radiation diagrams for antennas and graphs of ground-wave propagation for those trying to detect radio signals from satellites.

STSOrbit Plus (David H. Ransom, Jr., **rans7500@spacelink. nasa.gov**, **http://www.dransom.com**) is a DOS-based Earth satellite

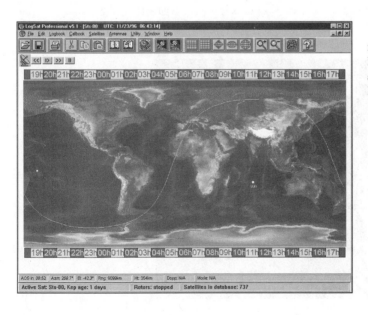

FIGURE 14.12
Plotting the orbit of a space shuttle in LogSat Professional

tracking application that is particularly targeted toward use during Space Shuttle missions. One primary satellite and up to 32 additional static or real-time satellites can be simultaneously tracked in real time on most computers. Both orthographic and rectangular VGA color map projections are available, displaying the Earth as a globe (Figure 14.13) or using the more traditional flat map. The display shows the selected satellite as a small symbol or icon, the projected orbital ground track for the next three hours and the past one and a half hours, and many other features including circles of visibility, TDRS coverage, and the solar terminator. STSOrbit Plus has operated on the Mir space station and on numerous Space Shuttle flights—even showing up on one of the big screens during a broadcast Space Shuttle flight. The application has many features that were implemented at the request of NASA astronauts.

Weather and Time Software

Unless you're strictly an armchair astronomer, weather is of great importance to your life in astronomy. Is it going to be clear tonight? Is wind forecast? How cold is it going to get? If you're

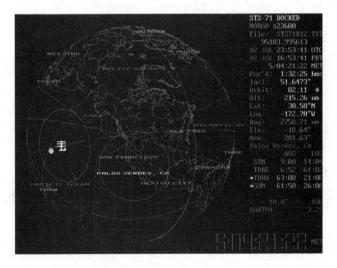

FIGURE 14.13
Tracking a shuttle mission in STSPlus

planning on spending time outside with your telescope, you'll need to know the answers to these questions. Fortunately, current local weather information is readily available for personal computers that have access to the Internet. Weather applications range from the fairly complex and sophisticated to the extremely simple—the needs of the amateur astronomer generally fall somewhere within the mid-to-lower range of these programs.

Similarly, time is of the essence, at least for the amateur astronomer. An accurate computer clock is important for everything from determining which objects are currently in the sky to making precise astronomical observations for scientific purposes. Computer clocks are generally not extremely accurate to begin with, and combine this with a multitasking operating system and you may find that your computer may deviate many minutes from accuracy. Although a computer clock that is off by a few minutes is generally not a problem for most users, for the amateur astronomer such an error can be significant. You might miss that instance of the Moon occulting Antares, for example, if you use software to determine when you should go outside and have a look. For scientific observations, of course, accurate timing is often critical. Whatever your particular needs, software is available that will access a standard time service site and adjust your computer clock to the correct time, even computing the amount of time required for the signal to traverse the telephone lines between your computer and the remote site.

WinWeather (Insanely Great Software, 800/319-5107, **igs@ igsnet.com**, **http://www.igsnet.com**) covers a broad range of weather needs. It is possible to get live pictures from both American and international weather cameras, hourly weather reports and forecasts, the latest satellite images, and such exotic items as the ultraviolet light forecast (Figure 14.14). This excellent application can track an unlimited list of cities of your choice, saving the data for review and evaluation even when you're not online. WinWeather is available as shareware for Windows 95, 98, and NT.

FIGURE 14.14
The variety of weather information that is available in WinWeather, from satellite images to forecasts to live Webcam photos

WetSock (Locutus Codeware, **info@locutuscodeware.com**, **http://www.locutuscodeware.com**) is another excellent weather application for use by amateur astronomers. WetSock shows your current weather conditions and forecasts as an icon on the tray notification area of the system taskbar, and it can show information on almost 3,000 cities worldwide (including 1,500 in the United States and 100 in Canada). WetSock operates as an icon on the Windows 95/98/NT taskbar, monitoring your dialup connections in the background and getting the weather information once you are connected for any purpose—checking your e-mail, for example. It can also dial in and hang up by itself either on command or at a user-selected interval.

AccuSet (Retsik Software, **retsik@aol.com**, **http://www.retsik-software.com**) has been available for some years now and boasts a wealth of sophisticated time functions. For example, AccuSet determines the accuracy of your PC's internal clock and can then make corrections to the system clock to compensate for the error rate—without the need for a telephone call (Figure 14.15). In addition to adjusting the PC's internal clock, AccuSet can also display the date and time for any five (user-configurable) cities around the world. It can also display the sunrise and sunset times for each

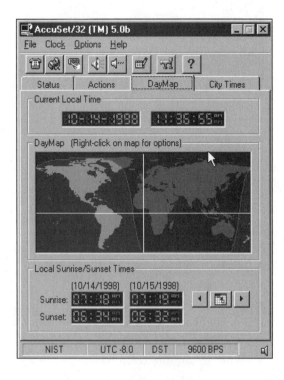

FIGURE 14.15
Checking the current local time using
AccuSet

specified location, and it automatically handles Daylight Saving Time. AccuSet is flexible, fast, and easy to use. A shareware version is available, and the registration fee is moderate.

GeoClock (GeoClock, 703/241-2661, **geoclock@compuserve. com, http://www.clark.net/pub/bblake/geoclock**) shows the current time (based on your computer's clock) with a map of the Earth. The current Sun position is displayed, and the parts of the Earth in sunlight and twilight are highlighted. This display is automatically updated every few seconds. Local sunrise and sunset, the Sun's azimuth and elevation, and times around the world are also displayed. A variety of map backgrounds and other options are available. A graphics adapter and hard disk are required. The DOS version of GeoClock is an EGA-VGA-SVGA program operating in 16-color mode. The GeoClock RBBS can be reached at 703/241-7980, with modem settings of N-8-1. GeoClock is distributed as shareware. With registration, you get the latest versions of the programs, 44 maps, zoom, distance measuring, local time displayed next to the city names, and immediate map display. GeoClock comes in DOS and Windows versions, and both are included with registration. The Windows version works with Windows 3.1, 95, 98, NT, and OS/2. The Windows version also includes screen saver and wallpaper modes.

SocketWatch (Locutus Codeware, **info@locutuscodeware.com**, **http://www.locutuscodeware.com**) can keep Windows 95, 98, and NT systems' time synchronized over the Internet. Like their application WetSock, discussed earlier, SocketWatch works in the background, monitoring your Internet connection and setting the PC clock when you check e-mail or browse those favored astronomy sites. Locutus Codeware updates SocketWatch fairly regularly, adding new features and stomping on a few bugs. A shareware version is available, and registration is reasonably priced. SocketWatch is certainly the most minimal of the time applications discussed here, appearing as an icon on the Windows taskbar. One handy feature of SocketWatch is its ability to sound an audible warning when it detects a severe weather alert from the National Weather Service, alerting you to pending inclement weather. For general use, you have only to right-click the SocketWatch icon to get the local weather forecast. This is a great application for the amateur astronomer who wants to have a quick check of the local weather.

Astronomical Image Processing Software

A picture might be worth a thousand words, but if it is impossible to clearly see what is trying to be communicated, then the vocabulary might be insufficient. While astronomical image processing software might seem to be strictly in the domain of those fortunate enough to have a CCD camera, in fact there are many uses for such software. For example, a broad selection of CD-ROMs containing raw image data from the Voyager flights through the outer Solar System are available, and image processing software can reveal an absolute wealth of information and allow even the novice to view previously unpublished images from this material. Other sources of raw astronomical image data exist on the Internet, and this material can be analyzed and processed by anyone willing to take the time to download the material. While several of the "Swiss Army Knife" planetarium software programs include limited image processing facilities, none offer the depth of features and capabilities of the stand-alone applications discussed here.

AstroArt (Fabio Cavicchio, **msb@ntt.it**, **http://www.sira.it/ msb/Astro_en.htm**) is a relatively new image-processing and analysis application for Windows 95, 98, and NT systems. One of its really nice features is the incorporation of the GSC catalog so that it is possible to perform astrometric and photometric calibration of images (Figure 14.16). AstroArt supports both the 8- and 16-bit FITS file formats produced by commercial CCD cameras and can import generic file formats of up to 4,000 by 4,000 pixels. A wide range of preprocessing and postprocessing filters are available for getting the most from an image, and a customizable macro function simplifies repetition of complex tasks. AstroArt also includes tricolor imaging capabilities, and it can be used as a blink comparitor to quickly shift back and forth between two images to look for

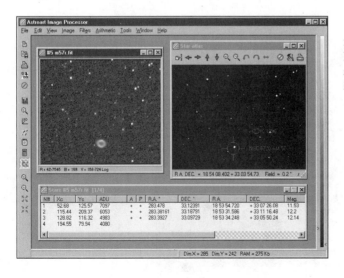

FIGURE 14.16
Identifying M57 and nearby stars in a CCD image using AstroArt

moving or changed objects. The full feature set is listed at the AstroArt Web site, and a demo version is available.

CCDSoft (Software Bisque, 303/278-4478, **http://www. bisque.com**), formerly known as SkyPro, combines both CCD camera control functions and astronomical image processing functions in one application. CCDSoft has virtually every feature you could imagine in such an image processing application, and then some (Figure 14.17). You can use CCDSoft in combination with TheSky (described earlier in this chapter) to control both CCD cameras and computer-driven telescopes—a powerful combination for the advanced amateur astronomer. Another great feature for those having both these applications is the ability to use TheSky in tandem with CCDSoft to easily identify celestial objects in your CCD images. CCDSoft lets you control everything from your CCD camera's focus to its resolution and temperature. The pre- and post-image processing functions are easily the most complete available in any astronomical image processing software targeted for the amateur market. For a full listing of the features of CCDSoft, check the Software Bisque Web site. If you're serious about astronomical imaging, this is one application you should investigate.

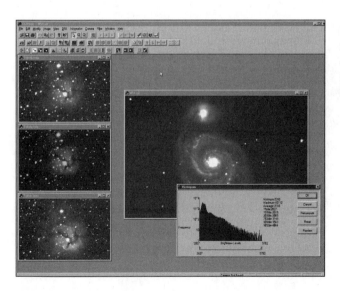

FIGURE 14.17
Analyzing CCD images of galaxies and nebulae in CCD-Soft

Educational and Other Software

Not all astronomy software is dedicated to providing views of the night sky to locate distant stars and galaxies. Many programs are available that provide information on the Solar System and the universe. These programs include educational packages that focus on a single topic in astronomy, tours of the Solar System and the universe with lots of images, commentary from famous scientists, and interactive software. Descriptions of some of the best programs are provided here.

Tours of Some Educational and Other Software Packages

A number of programs are dedicated to exploring the various worlds of our Solar System. **Nine Worlds** (Palladium Interactive, 415/464-5500, **webstar@palladiumnet.com**, **http://www.palladiumnet.com**) is a tour of the nine planets of the Solar System, highlighted by narration from actor Patrick Stewart (*Star Trek*'s Captain Picard; see Figure 14.18). The program includes detailed information about each world, with plenty of images and video clips; and a link to a special Web site where users can find out

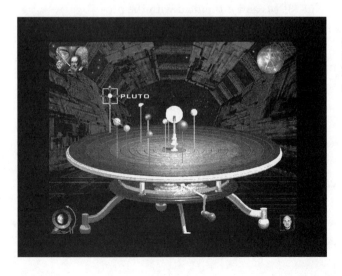

FIGURE 14.18
Select the planet you want to learn more about from this screen in Nine World

even more about the Solar System and jump to other Web sites. Nine Worlds is available on CD-ROM for Macintosh and Windows computers.

Beyond Planet Earth (Discovery Channel Interactive, 301/986-1999, **product_info@discovery.com**, **http://multimedia. discovery.com/mms/beyond/beyond.html**) explores several aspects of the Solar System. The CD-ROM includes hundreds of images, video clips, and other information about the Solar System and what lies beyond. It includes a section on what a human mission to Mars would be like and a series of interviews with space experts, including former astronauts Buzz Aldrin and Kathryn Sullivan. Beyond Planet Earth is available for Macintosh and Windows computers.

The Planets (Scientific American Library/Byron Preiss Multimedia, 800/945-3155, **welcome@bpmc.com**, **http://st2.yahoo.com/ byronpreissscienamunpla.html**) is billed as "the ultimate tour of the Solar System," and it certainly comes close. The CD-ROM features a gallery of images and movies, a guided tour through the Solar System, and 3D flybys of the planets. If you're not satisfied with our solar system, you can create your own in the program and then have astronomer Dr. Donald Goldsmith provide commentary on your new worlds (see Figure 14.19). It's available on separate Windows and Macintosh CD-ROMs and is bundled with the program The Universe (described later).

If your interest lies in exploring the Solar System, you will find **Planetary Missions** (formerly **Solar System Explorer**) (Piranha Interactive, 602/491-0500, **sales@piranhainteractive.com**, **http:// www.planetarymissions.com/**) an intriguing program. It allows you to design and fly missions to bodies throughout the Solar System using an advanced twenty-first-century spacecraft, which helps keep the time spent traveling from world to world to a minimum (see Figure 14.20). You can then study each world from orbit and send probes to the surface to conduct further studies and return samples to your spacecraft before continuing your mission. Planetary Missions is available on CD-ROM for Macintosh and Windows machines.

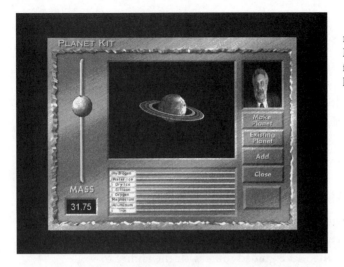

FIGURE 14.19
Make your own planets and
find out what they would be
like in The Planets

FIGURE 14.20
Plan missions to other worlds
in Planetary Missions

A Field Trip to the Sky (Sunburst Communications, 914/747-3310, service@nysunburst.com, http://www.nysunburst.com) is designed as an educational exploration of the Solar System for middle-school students. It includes interactive demonstrations of the relationships among Earth, the Sun, and the Moon, and it includes detailed information about each planet in the Solar System. The program is available on CD-ROM for Macintosh and Windows systems; discounts are available for multiple copies for use in schools.

For a detailed view of the surface of Venus, check out **Venus Explorer** (RomTech, 805/781-2250, **http://www.romt.com**). This program provides an interface to high-resolution radar images of the planet's surface, obtained by the Magellan spacecraft in the early 1990s. You can zoom in on specific regions or search for areas by name, although you won't find much other information about Venus or about what the images show. The program is available on CD-ROM for Macintosh and Windows systems.

The same company also makes **Mars Rover**, which gives users a 3D view of Mars's surface. You can choose from a number of different areas to explore and then take guided tours with the ability to zoom in and out. The program uses data from the Viking missions to Mars in the 1970s to provide accurate depictions of the Martian landscape. Mars Rover is available for Macintosh and Windows systems, bundled with Distant Suns (described earlier).

There is also no shortage of programs devoted to astronomical phenomena outside our Solar System. **Discover Astronomy** (Maris Multimedia, 415/492-2819, **redshift@maris.com**, **http://www. maris.com/**) provides an introduction to a wide range of topics, from the constellations to the origin of the universe. The program features some of the narrated animation found in its sibling program, Redshift 2 (described earlier), but it also has some of its own animation and many images of astronomical phenomena (see Figure 14.21). Discover Astronomy is available on CD-ROM for Windows.

Galaxy Guide (Parallax Multimedia, 916/887-0225, **parallax@ quiknet.com**, **http://www.quiknet.com/~parallax**) is a wide-ranging introduction to astronomy outside the Solar System. There are detailed maps of constellations and explanations of a broad range of topics, complete with many images. There's also a game to test your knowledge, with several levels of difficulty. The program is available on CD-ROM for Windows.

FIGURE 14.21
Learn about the constellation Orion in Discover Astronomy

The Universe (Scientific American Library/Byron Preiss Multimedia, 800/945-3155, **welcome@bpmc.com**, **http:// st2.yahoo. com/byronpreissscienamunpla.html**) is based on the Scientific American Library book *From Quarks to the Cosmos* and explores a wide range of phenomena, from the smallest subatomic particles to the origin of the universe. The program includes interviews with leading physicists and astronomers; narrated information on leading topics in astrophysics; and interactive experiments to study the evolution of stars, galaxies, and the universe. The Universe is on CD-ROM for Macintosh and Windows and is bundled with The Planets (described earlier).

Star Probe (Artemis Science Curriculum and Software, 503/284-7446, **artemis@teleport.com**, **http://www.teleport.com/ ~artemis**) is an educational program designed to teach students how stars work. It lets you study one of several types of stars, from our own Sun to red giants like Betelgeuse (see Figure 14.22). You can explore the star from a distance or probe deep within it to see what conditions are like. The program is available for Macintosh computers (a Windows version is in development).

Binary Star (Gemini Software, 501/442-4159, **clacy@comp. uark.edu**) allows students to study the characteristics of binary star

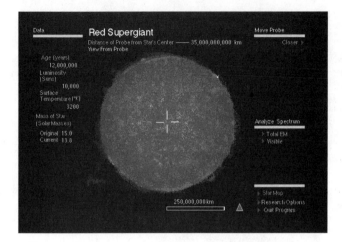

FIGURE 14.22
Star Probe offers a close-up view of the red giant star Betelgeuse

systems. Users create their own binary star systems and see how these star systems would look when observed from Earth. Binary Star is written for Macintosh computers.

H-R Calc 2.0 (David Irizarry; P.O. Box 77243, Seattle, WA 98177; **xerxees@ix.netcom.com**) is a shareware program that explores the H-R diagram, the relationship between the brightness and the spectral class of a star that helps define the different classes of stars. By choosing a particular location on the diagram, you can find out some basic characteristics of a type of star, including its size and surface temperature. You can also manually alter these values to see how they affect the location of the star on the diagram. The program is available for Windows.

Other Educational and Astronomy Software

Other educational software packages about astronomy are available, far more than we can list here, as well as a number of miscellaneous astronomy packages. A sampling is provided here.

Astronomy Lab and **Astronomy Clock** (Personal MicroCosms, 303/753-3268, **EricTerrell@juno.com**, **http://members.aol.com/ericb98398/index.html**) are Windows programs that illustrate basic aspects of astronomy. Astronomy Lab demonstrates basic concepts in astronomy and provides predictions of eclipses and rise and set times for astronomical objects. Astronomy Clock explains the different time systems used in astronomy. Both programs are available in separate versions for Windows 3.1 and 95/NT.

Earthwatch (Elanware, Inc., 134 Normandy Dr., Brunswick, OH 44212) is a shareware program that provides a world map showing where it's currently daytime on Earth. The map also shows the time zones and provides information on Moon phases. The program is written for DOS, but is usable under Windows, with a registration fee.

Exploration in Education (Space Telescope Science Institute, **ExInEd@stsci.edu, http://www.stsci.edu/exined-html/exined-home.html**) is a collection of educational programs about astronomy, and specifically the Solar System, created by the Space Telescope Science Institute (the agency that operates the Hubble Space Telescope for NASA). Topics include Venus, Mars, the Apollo missions to the Moon, and some of the best images from the Hubble Space Telescope. The software is available for Macintosh and Windows computers and can be downloaded from the Web site at no charge.

Explore the Planets (Tasa Graphic Arts, 800/293-2725, **tasagraph@aol.com, http://www.swcp.com/~tasa/progexplore.html**) is an exploration of the Solar System, with an emphasis on comparative planetology: examining what we know about other worlds to learn more about our own planet. The program specifically targets high school students. The CD-ROM is available for Macintosh and Windows computers.

JupSat95 (Gary Nugent, **gnugent@indigo.ie, http://indigo.ie/~gnugent/JupSat95**), as you might gather from the name, is a small application for Windows 95, 98, and NT that calculates and displays the positions of Jupiter's four main (Galilean) moons for any date and time. JupSat95 offers both a plan view (from above), showing the current moon positions, as well as a side view that illustrates how the moons currently appear through binoculars or through an astronomical telescope or the star diagonal on such a telescope (Figure 14.23). Among other features, JupSat95 can display and print satellite tracks for a specified month and year; display and print the Great Red Spot transit time, also for a specified month and year; and animate a simulation of the Jovian system using a user-specified interval. This application is quite useful for identifying which of the moons are currently visible and is very nicely designed.

MoonPhase (Locutus Codeware, **info@locutuscodeware.com**, **http://www.locutuscodeware.com**) displays the current phase of the Moon on the tray notification area of the system taskbar on systems using Windows 95 and up or Windows NT4 and up. With a glance at your taskbar, you can now know the current phase of the Moon and plan any observation sessions accordingly.

On Top of the World (Exploration Software, **hlynka@tiac.net**, **http://www.tiac.net/users/hlynka**) gives users a new perspective on Earth, letting you view a 3D model of Earth from above and explore thousands of named features. The program runs on Windows and has a registration fee.

Planet Guide (Maraj, Inc., 770/413-5838, **maraj@bigfoot.com**, **http://www2.gsu.edu/~usgnrmx/maraj/products/prod01.html**) is an astronomy tool that provides a brief amount of information on all the planets in our Solar System. This handy and informative application can provide answers to numerous questions about the planets circling our Sun, on topics ranging from the composition of planetary atmospheres to how much you would weigh on the surfaces of different worlds (Figure 14.24).

Project CLEA (Gettysburg College, 717/337-6028, **clea@gettysburg.edu**, **http://www.gettysburg.edu/project/physics/clea/CLEAhome.html**), Contemporary Laboratory Experiences in Astronomy, offers several educational software packages designed for use in schools and colleges. Programs include studies of the moons of Jupiter, craters on our Moon, the expansion of the universe, and more. The software is written for Macintosh and Windows computers and is free.

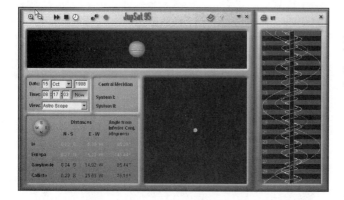

FIGURE 14.23
Various views of the Galilean moons relative to Jupiter, as shown in JupSat95

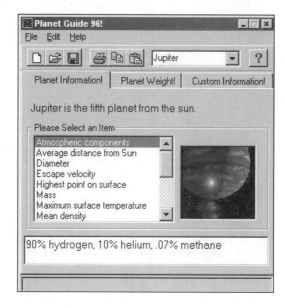

FIGURE 14.24
Check various facts about Jupiter or other planets with Planet Guide

Besides this collection of software packages about all aspects of astronomy, there is a wide range of online resources on the subject. The next chapter discusses some of the online resources (besides Web sites) that you can use to learn more about astronomy.

15

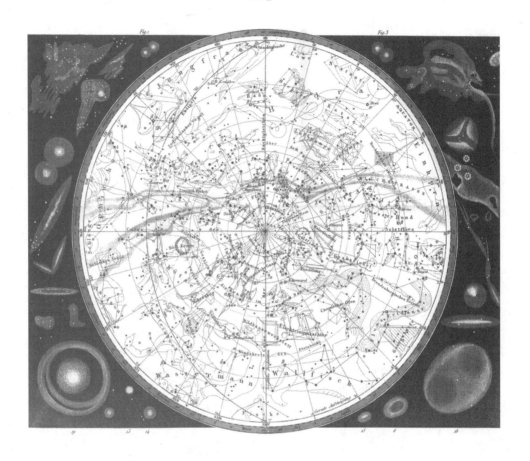

EXPLORATIONS ON YOUR OWN: ONLINE

Throughout this book we have been emphasizing Web sites as an online resource for more information on particular topics. While there is a tremendous amount of information available on the Web today, it's not the only source of online information. Usenet newsgroups provide people with the opportunity to discuss and debate issues in astronomy and related fields. Mailing lists serve a similar purpose, while chat rooms often allow people to discuss astronomy with experts in the field. Outside the Internet, online services like America Online and Prodigy offer their own resources to learn more about astronomy and share it with others.

This chapter will focus on online astronomy-related resources other than Web sites. This includes online services, newsgroups, mailing lists, and discussion areas. We won't entirely ignore the Web, though; we'll also show how to find the resources you're looking there. See Appendix D for a comprehensive list of astronomy Web sites.

Online Services

Long before the Web, or the Internet in general, became popular, online services like America Online, Prodigy, and CompuServe provided subscribers with some of the best astronomy resources available online. While much of the best content has migrated to the much larger audience of the Web, many of the online services still maintain separate sections, which often include information and features you won't find as easily on the Internet.

To access these forums, though, you have to subscribe to these online services.

America Online, the largest of these services, offers a wide range of information and other features related to astronomy and space. The Astronomy Club (AOL keyword **astronomy**) is maintained by Mr. Astro, who in real life is Stuart Goldman, an editor for *Sky and Telescope* magazine (see Figure 15.1). The section features a news and information center where the latest events in astronomy are posted. Chat sessions are held in the Planetarium, which is also available for general astronomy discussions at times other than the scheduled weekly chats. Message boards provide another way for users to interact by exchanging messages on a variety of astronomical topics. The section has a library of downloadable astronomy programs and a collection of links to astronomy resources, including Web sites.

You can find a similar set of resources, with an emphasis on the planets and space exploration, at Space Exploration Online (AOL keyword **space**), operated by volunteers from the National Space Society (see Figure 15.2). The emphasis here is on space exploration, as evidenced by the sections on the Space Shuttle and the Russian space station Mir. However, those interested in the planets will want to check out the Planet Probes section, which includes

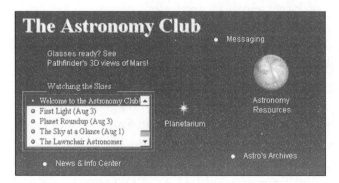

FIGURE 15.1
The Astronomy Club offers a number of resources related to astronomy for America Online members

FIGURE 15.2
Space exploration,
including planetary
missions, is the
focus of Space
Exploration Online
on America Online

information and links to other resources about spacecraft missions throughout the solar system. The section also hosted a number of chats about Mars during the Mars Pathfinder landing. The section includes Space Talk message boards, a large library of information and images about space exploration and planetary missions, and Ask an Astronaut, an opportunity for members to ask questions of current and former astronauts.

CompuServe, another major online service, also provides considerable astronomy and space resources. A CompuServe user can start with a visit to the Astronomy and Space section of Science Sphere for basic information (keyword **GO ASTRONOMY**; see Figure 15.3). The information available here is limited: a handful of links to astronomy Web sites such as The Nine Planets and some local resources, such as sunspot numbers and collections of NASA and ESA (European Space Agency) press releases.

The Science Sphere section, though, includes links to CompuServe's astronomy and space forums, which provide the best information and services available on the online service. The Astronomy Forum (accessible from keyword **GO ASTROFORUM**) includes message boards where people can read and write messages on a variety of topics related to astronomy, from the latest news to tips for amateur astronomers (see Figure 15.4). The forum also includes file libraries where users can download images and programs; and conferences, where users can discuss astronomy topics. A highlight of the forum's conference room is a weekly discussion held Sunday afternoons, moderated by Kelly Beatty, senior editor of *Sky and Telescope* magazine.

Similar features are available in two related forums. The Space

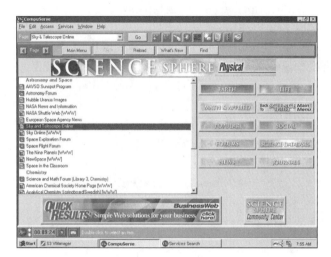

FIGURE 15.3
The Science Sphere section has links to basic information available on both CompuServe and the Web

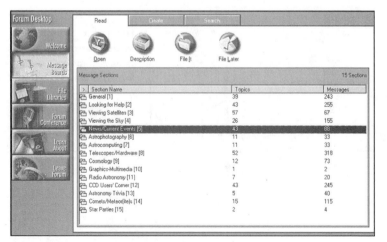

FIGURE 15.4
The message board in the Astronomy Forum on CompuServe provides users with a wide variety of topics

Exploration Forum (**GO SPACEX**) features message boards, file libraries, and conference areas that deal with the exploration of space, primarily missions to other planets in the Solar System. The recent success of the Mars Pathfinder mission has resulted in a spike of interest in Mars in this forum. Manned space activities, from the Shuttle and Mir to plans for missions to the Moon and Mars, are discussed in the Space Flight Forum (**GO SPACEFLIGHT**), which has the same features as the other forums.

Prodigy, with a much smaller membership base than America Online, also offers fewer astronomy resources to its members. There is no specific "community," in Prodigy's terminology, about astronomy. The closest community is the Space and Flight Pad, which includes information about aviation and space flight. One subsection of the community is titled Great Astronomy Images and Info (Figure 15.5), but an examination of this section shows it to be just a collection of links to sites on the Web, something that's easily available to anyone with Web access, regardless if you're a member of Prodigy or not. Since more extensive collections of astronomy Web sites are available for free on the Web (or from this book!) there is little reason to use Prodigy to find exclusive astronomy resources.

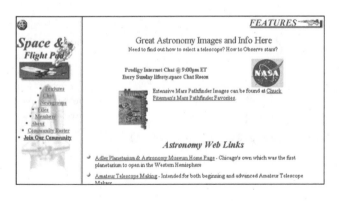

FIGURE 15.5
The only astronomy section on Prodigy turns out to be a collection of links to Web sites

Usenet Newsgroups

Long before the Web and the Internet became commonplace, one popular online resource was Usenet, a collection of discussion areas distributed among Internet computers (or among those who could dial up a computer connected to the news feed). You can discuss anything and everything on Usenet, and a number of Usenet newsgroups are devoted to astronomy.

To access Usenet, you need a program generally called a newsreader to read the messages in the newsgroups and to post your own messages. A newsreader is built into Netscape Navigator and Communicator and is included as a separate program distributed with Microsoft Internet Explorer, so if you have a Web browser you probably already have a newsreader (see Figure 15.6). If you don't, or you would prefer a different program, check the list of newsreaders at **http://www.yahoo.com/Computers_and_Internet/Software/Internet/Usenet.**

Several newsgroups are devoted to astronomy discussions. Each has its own purpose, and the quality and quantity of postings varies widely. What follows is a primer on the good, bad, and ugly sides of the astronomy newsgroups.

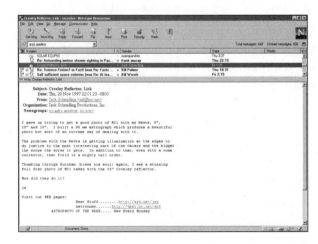

FIGURE 15.6
The newsreader built into Netscape Communicator

The main newsgroup is **sci.astro**. For many years this was the sole location on Usenet for astronomy discussions, but as the volume of messages rose different areas split off into their own groups, described below. However, quality has not kept up with quantity. On a typical day more than 100 messages may be posted to the newsgroup, but only a small fraction will be relevant to astronomy. Many of the rest are long-running arguments, tangents into religious or philosophical discussions, or unsolicited commercial messages (often called *spams*). To get the most out of sci.astro you need to take advantage of your newsreader's ability to select only the discussions, or threads, of interest to you. You may also be interested in your newsreader's ability to create *killfiles*, which are lists of discussion topics or people from which you don't want to see messages. Both techniques allow you to read the messages you find interesting while ignoring the rest. Volunteers maintain a list of frequently asked questions, with answers, to try to keep the same questions from being asked again and again in the newsgroup. This FAQ is posted regularly to the newsgroup and is available at **http://sciastro.astronomy.net**.

One newsgroup that attracts as much traffic as sci.astro is **sci.astro.amateur**. As the name suggests, the topic of this newsgroup is amateur astronomy. Amateurs from around the world share their experiences observing; ask questions about the best telescopes, binoculars, and other equipment to use; and more. Although this newsgroup gets as many postings in a day as sci.astro, far more of the messages in sci.astro.amateur are relevant to the topic. If you are interested in amateur astronomy in any way, you'll find a wealth of information in this newsgroup. The FAQ for sci.astro.amateur is on the Web at **http://www.xmission.com/~dnash/astrodir/saafaq**.

The newsgroup **sci.astro.research** is devoted to specific research topics in astronomy. The newsgroup is moderated, which means any postings to the newsgroup first have to be cleared by a volunteer moderator, who screens out inappropriate questions to keep the content of the newsgroup relevant. On a typical day only a few messages are posted to the newsgroup. More information about sci.astro.research is at **http://xanth.msfc.nasa.gov/xray/sar.html**.

If you're interested in planetariums, you'll want to check out **sci.astro.planetarium**. Employees and volunteers of planetariums discuss presentations, educational programs, and related issues. More information on the newsgroup, including a FAQ, is on the Web at **http://www.lochness.com/pltref/sapfaq.html**.

The newsgroup **sci.astro.hubble** is not a discussion forum; no postings are allowed by other than authorized individuals, and no discussion takes place. Instead, information about the status of the Hubble Space Telescope is posted here on a daily basis. You can read what observing programs the telescope is performing on a given day and what is planned for the near future. Press releases about discoveries made with Hubble are also posted here. More information about the newsgroup is at **http://tycho.la.asu.edu/sah.html**.

Research astronomers and others involved in the detailed analysis of astronomical data may be interested in several low-volume newsgroups. The newsgroup **sci.astro.fits** is for questions and discussion about the FITS (Flexible Image Transport System) format, the standard format for astronomical data that professionals use. Information about FITS in general is at **http://www.cv.nrao.edu/fits**. Two other newsgroups, **alt.sci.astro.aips** and **alt.sci.astro.figaro**, are devoted to specific applications some astronomers use, AIPS and Figaro. The Web has information about these software packages at **http://www.cv.nrao.edu/aips/whatisaips.html** and **http://asds.stsci.edu/asds/packages/analysis/FIGARO.html**, respectively.

People interested specifically in planetary science might think the **alt.sci.planetary** newsgroup would be of interest. However, only a dozen or two postings are made to this newsgroup on a given weekday, and most of those aren't relevant to planetary science at all. More planetary science discussion takes place in **sci.astro** or **sci.space.science** (see below).

There is a separate hierarchy of space newsgroups that often contain topics relevant to astronomy. The most relevant is **sci.space.science**, a moderated newsgroup for the discussion of space science topics, primarily planetary science. This newsgroup gets a few dozen messages on a typical day, and since the newsgroup is moderated there's not much noise or irrelevant messages. Other space newsgroups include **sci.space.tech** (a moderated newsgroup for discussion of space technology), **sci.space.policy** (unmoderated discussion of space policy), **sci.space.history** (discussion of the history of space exploration), **sci.space.shuttle** (discussion of the Space Shuttle), and **sci.space.news** (a tightly moderated newsgroup for press releases and other information related to space and astronomy).

Mailing Lists

Another method for astronomy enthusiasts to communicate is through astronomy mailing lists. In a mailing list, messages sent to a specific address are then forwarded to people who have subscribed to that list. One advantage of mailing lists is that they require no special software other than whatever you currently use for e-mail. It is an advantage, though, to have e-mail software that can sort your mail so all the messages from a single list are placed in a separate folder; then you can read them as a single group instead of mixed with mail from other lists or personal messages.

There is a small number of mailing lists on astronomy available. Those interested in building their own telescopes, including the long process of grinding and polishing mirrors and building mounts, will want to check out the Amateur Telescope Makers Mailing List. This list exists to allow amateur telescope makers to share experiences and ask questions about telescope making. To subscribe, send an e-mail message to **majordomo@shore.net**. In the body (not the subject) of the message, type **subscribe atm**. You'll then receive a welcome message with more information about posting messages to and unsubscribing from the mailing list.

For those with a more historical bent, the HASTRO-L mailing list on the history of astronomy may prove appealing. It sets no limits to the discussion of the history of astronomy; all cultures, all time periods, and all approaches to the subject are accepted. To subscribe, send a message to **listserv@wvnvm.wvnet.edu** with the line **subscribe hastro-l** in the body.

In addition to these discussion-oriented mailing lists, several mailing lists distribute information about astronomy from a single source, but are not open for discussion. A mailing list related to the radio program *Earth and Sky* provides information about the astronomy topics discussed in each show. To subscribe, send e-mail to **earthandsky-request@earthsky.com** with the word **SUBSCRIBE** (in capital letters) in the body of the message. More information about the mailing list and the radio program is at **http://www.earthsky.com**.

If you're interested in receiving press releases and other information related to astronomy and space, check out the SEDSNEWS-L mailing list. Sponsored by Students for the Exploration and Development of Space (SEDS), it includes press releases from NASA and ESA as well as other news items. To subscribe, send e-mail to **listserv@tamvm1.tamu.edu** and the words **subscribe sedsnews-l** in the body of the message.

For more news about astronomy and space, along with in-depth articles and other features, read *SpaceViews*. A publication of the Boston chapter of the National Space Society, the newsletter reaches several thousand people a month in dozens of countries worldwide. Published twice a month, each issue includes the latest news in space exploration and astronomy, as well as articles, book and Web site reviews, and other information. To subscribe to the *SpaceViews* mailing list, send e-mail to **majordomo@spaceviews.com** and write **subscribe spaceviews** in the body of the message. You can also subscribe from the SpaceViews Web site at **http://www.spaceviews.com**.

Other mailing lists related to astronomy include NightSights, a daily newsletter about what's visible in the night sky. You can subscribe to the newsletter as well as view current and past editions at **http://www.bignetwork.com/dp/ns/**. The astro-photo mailing list is a discussion list devoted to astrophotography. To subscribe, send mail to **majordomo@nightsky.com** with **subscribe astro-photo** in the message body. Another list, radio-astronomy, discusses the field of radio astronomy. Subscribe by sending mail to **majordomo@qth.net** with **subscribe radio-astronomy** in the message body. To keep track of these and other astronomy mailing lists, visit the astronomy section of Liszt, an online database of over 90,000 mailing lists, at **http://www.liszt.com**, in particular its "select" list of recommended astronomy groups, at **http://www. liszt.com/select/Science/Astronomy/**.

Discussions and Chats

One disadvantage of Usenet newsgroups and Internet mailing lists is the lack of immediate feedback. You can post a question or comment to a newsgroup and wait hours or days for a response, if any is forthcoming. Newsgroups and mailing lists serve their purposes, but if you're looking to communicate and interact with other astronomy enthusiasts, from around the corner or around the world, you will want to investigate chats and online discussions.

Foremost among online discussion media is Internet Relay Chat, or IRC. This method of discussion requires logging onto one of a networked group of servers using IRC software. You can then go to any number of discussion areas, or channels, and talk in real time with other people who are online at the time. For more information on IRC, including how to find the appropriate software for your computer, visit **http://www.connected-media.com/IRC**.

There are two main discussion channels on IRC for astronomy, **#astronomy** and **#sciastro**. There is little difference between the two; both offer general discussion of astronomy topics and are open 24 hours a day, so you can usually find someone on one or both channels at any time. There are also scheduled chats at specific times to try to bring together as large a group as possible at a single time. Both channels have a group of regulars who appear often on the channels, but newcomers are also welcome. For more information about the channel #astronomy, visit **http://www.sci.fi/~riku/astronomy**. Check **http://sciastro.net/chat** for details on the #sciastro channel.

There are alternatives to IRC discussions, thanks to the growing use of chat programs, many written in Java, on Web sites. It can be hard to keep track of these chats, since some do not take place regularly or are special events. To help keep track of astronomy chats on Web sites around the world, visit **http://www.yack.com**, which keeps a *TV Guide*–like listing of each day's chats, and **http://events.yahoo.com**, which also keeps track of special and regularly scheduled chats online.

Finding It on the Web

Even with all the alternatives listed above, the Web is still the most popular source of information. However, as the Web continues to grow, finding information on astronomy-related topics becomes more challenging. To make the best use of your time online, you need to be able to find what you're looking for as quickly as possible.

One resource, of course, is the comprehensive listing of Web sites in the appendix of this book. However, Web site URLs are ethereal; sites can move or just disappear, making the old address invalid. Moreover, new sites appear constantly, so a static list loses value over time. When a list like the one in the appendix fails, you need to be able to search out the information online yourself.

One option is the search engines that keep track of Web sites everywhere. Some of the best include AltaVista (**http://altavista.com**), Infoseek (**http://www.infoseek.com**), and Lycos (**http://www.lycos.com**). These sites keep track of tens of millions of Web sites around the world, so when you conduct a search using one of these search engines, keep your search terms as specific as possible to limit the number of sites it finds. You can do a search on just *astronomy* and discover the search engine has found hundreds of thousands of sites, but that's not much help when you're looking for a specific topic.

Another option is a hierarchical listing of Web sites, organized by subject. One example is Yahoo; you can find its list of astronomy Web sites at **http.//dir.yahoo.com/Science/Astronomy**. The sites are divided into subtopics, making it easier to find the particular area that suits you.

There are also sites that specialize in listing astronomy Web sites. Astronomy Net (**http://www.astronomy.net**) features a searchable database of a small but growing number of astronomy Web sites. The Expanding Universe (**http://www.mtrl.toronto. on.ca/centres/bsd/astronomy**) uses the Dewey decimal system to categorize its collection of links, as though they were books in a library (this should be no surprise, since this site is hosted by the public library in Toronto.) Astroweb (**http://www.stsci.edu/ astroweb/astronomy.html**) has a list of hundreds of Web sites, divided by topic, with an emphasis on research and professional sites, as well as the home pages of professional astronomers and students. Amateur astronomer Mike Boschat maintains an impressive list of links for all aspects of astronomy at his Web site, **http://www.atm.dal.ca/~andromed**. Spacesites (**http://www. spacesites.com**) promises to provide a database of "space fact and fiction" sites on the Web, including sections on astronomy and planetary science, and should be operational by the time you read this. Yahoo keeps a list of these and other sites that maintain lists of astronomy links at **http://www.yahoo.com/Science/Astronomy/ Web_Directories**.

So where do you go from here? That is entirely up to you. This book has given you the tools for exploring the universe of information about astronomy available for computers both online and in software packages. To get the most out of your computer, you will need to study your resources and find ways to acquire new resources online or elsewhere. And, like generations of astronomers before you, discover.

16

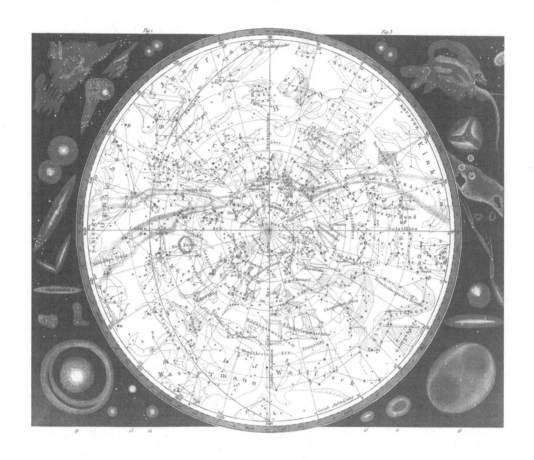

REMOTE ASTRONOMY
OVER THE INTERNET

The classical picture of an astronomer, one that comes to mind to most people even today, is that of a man hunched over the eyepiece of a telescope, either outdoors or inside a dome, peering into the night sky. This image has been etched into our collective consciousness through generations of repetition in books, movies, and television shows.

While that image might once have accurately portrayed how professional astronomers worked—and it is how many amateur astronomers observe today—it is terribly outdated. Today's telescopes are computer-controlled, and their instruments are also electronic, allowing professional astronomers today to work in the comfort of "warm rooms" at observatories, conducting their observations and analyzing their data from computers with rarely any need to venture out to the telescope itself, let alone look through an eyepiece.

While most professional astronomers today do at least go to the observatories where the telescopes are located to make their observations (unless, of course, their telescope is located in space, like the Hubble Space Telescope!), the rapid growth of the Internet is making even this less important. Many telescopes today can be operated remotely, allowing astronomers to work from their own offices thousands of kilometers away. Some telescopes have been set up to operate remotely over the Internet for not only professional astronomers, but amateurs, students, and anyone else interesting in capturing an image of a particular region of the night sky. In this chapter we take a look at these remotely-operated telescopes accessible to users over the Internet.

Publicly Accessible Automated Telescopes

A number of automated telescopes are publicly accessible on the Internet. With these telescopes you can submit requests for observations—usually after registering with the system, at no charge—and check on the status of the telescope. In most cases the requests for observations are paced in a queue, so you won't get your image immediately, but have to wait days, weeks, or even months, depending on the popularity of the telescope and the weather at its site, to get your images.

Bradford Robotic Telescope

The grandfather of all Internet-accessible robotic telescopes is the Bradford Robotic Telescope (**http://www.telescope.org/rti/**) (Figure 16.1). Located in West Yorkshire, England, the 46-cm (18-inch) telescope has been accessible to anyone on the Internet since 1993. It has been the most widely used of all the automated telescopes on the Internet.

To use the telescope, individuals have to register (at no charge) and obtain a password, all of which is handled automatically at the Web site. As of June 1999, there were over 60,000 registered telescope users from over 100 countries! Once registered, users can

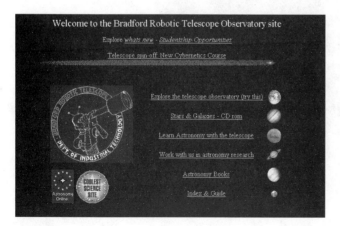

FIGURE 16.1
The Bradford Robotic Telescope Web site

submit observing requests for planets, stars, galaxies, and other objects in the night sky visible from the telescope site (Figure 16.2).

The observing requests are then placed into a database and automatically scheduled and prioritized. Each night, the telescope operates, the computers generate a list of observations for the telescope to perform. At the end of the night the images are returned and made available to those who requested them. Given the high demand for telescope time (and the notoriously poor English weather), it can take some time for an observing request to be fulfilled.

These delays have been exacerbated by technical problems with the telescope, which have taken it offline for months at a time while the staff at the University of Bradford repair the system. The Bradford group has also been working on a similar system on Tenerife, in the Canary Islands (**http://www.telescope.org/ tenerife/**), that has kept them from devoting their full attention to the original telescope. Thus, it could take months for an observing request submitted now to be fulfilled; at last check, there were over 5,000 observing requests in the queue!

FIGURE 16.2
An image of the Moon taken with the Bradford Robotic Telescope, a sample of the telescope's capabilities

University of Iowa Automated Telescope

Fortunately, the Bradford telescope is not the only telescope available for use by individuals over the Internet. The University of Iowa has set up a small (7-inch) telescope on the roof of one their buildings and allow people to use it over the Internet. The University of Iowa Automated Telescope Facility is used mainly by students in astronomy classes at the university but is also open to other students, teachers, and anyone else with an interest in astronomy. Last year it was used by more than 1,000 people, who took more than 15,000 images with the telescope.

To use the telescope, you first have to fill out a short form on the Web site (**http://inferno.physics.uiowa.edu**). Once the form is processed (which may take several days), you'll be e-mailed a remote observers code for use of the telescope. With that code, you can submit observing requests using a form at the Web site, with the limitation that no one observer is allowed to use the telescope for more than 40 minutes a night. The images are then made available over the Web site (Figure 16.3).

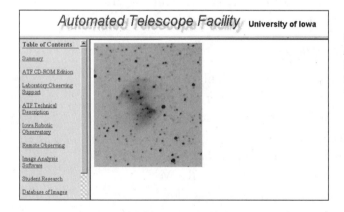

FIGURE 16.3
An image of M27, the Dumbbell Nebula, taken from the University of Iowa's Automated Telescope Facility

Building on the success of the current system, the university, in conjunction with Iowa State University and the University of Northern Iowa, is building a 50-cm (20-inch) remotely-operated telescope near Tucson, Arizona. The Iowa Robotic Observatory (**http://inferno.physics.uiowa.edu/IRO/iro_index.shtml**) will be primarily used by students, but about 20 percent of the time on the telescope will be made available to general Internet observers.

Remote Access Astronomy Project

The Remote Access Astronomy Project (**http://www.deepspace. ucsb.edu/rot.htm**) is similar to the University of Iowa project above. The project, started in 1992, features a 14-inch telescope mounted on roof of a building on the campus of the University of California at Santa Barbara. Image requests for the telescope can be submitted through a BBS (805-893-2650) or their Web site.

No registration is required to submit the image request. The emphasis of the telescope is on educational projects at the university and other schools; however, the program does accept a limited number of image requests from amateur astronomers on the Internet. A gallery of images of the telescope, and images taken with the telescope, can also be checked out from the Web site (Figure 16.4).

Eyes on the Skies Solar Telescope

Telescopes that view the night sky are not the only ones available on the Internet. The "Eyes on the Skies" observatory, located in Livermore, California, east of the San Francisco Bay area, provides images of the Sun on clear days to anyone on the Internet at **http:// www.hooked.net/~tvs/eyes**. The telescope has been accessible for a number of years, first as a dial-in BBS, and now on the Web.

The site features a sophisticated interface that allows visitors a great deal of control over the image of the Sun they see (Figure 16.5). Observers can pan across the disk of the Sun, focus in on certain regions, and tweak the image quality. Old images and movies of the Sun are also available from the site, if it's cloudy or dark.

FIGURE 16.4
An image of the
M20 nebula from
the Remote Access
Astronomy Project

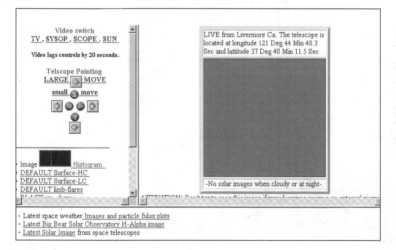

FIGURE 16.5
The Web interface
for controlling the
Eyes on the Skies
solar telescope
provides users
with a wide array
of options

Nassau Station Robotic Telescope

The latest robotic telescope to go online is the Nassau Station
Robotic Telescope (**http://astrwww.cwru.edu/nassau/nassau.html**),

located 30 miles east of Cleveland and operated by the Case Western Reserve University (Figure 16.6). The 0.9-meter (36-inch) telescope was in the process of getting online as of mid-1999, and should be fully up and running by the time you read this. The telescope operation is completely automated; if a camera trained on Polaris shows clear skies for at least 30 minutes, the telescope dome opens and observations are conducted. Sensors check for high humdity, rain, snow, or high winds, and instruct the dome to close if any of those hazardous conditions are detected.

The telescope is open to the public to submit observations. New users must first register for a username and password (available for free), and then are able to submit "proposals" on a page of the telescope's Web site for observations. These proposals must specify the object you want to observe (either by name or location in the sky), the number of exposures, the length of time for each exposure, and what filters, if any, should be used. A standard set of U, B, V, R, and I filters are available. Users can select up to five exposures per proposal. The completed proposal is then submitted to the telescope, which places it in its database. The telescope then carries out the observations based on if and when the object is observable, the amount of time it will take to complete the observations, how long the proposal was been waiting in the "queue," and

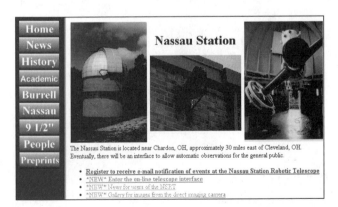

FIGURE 16.6
The Web site for the Nassau Station Robotic Telescope in Ohio

whether the proposal is from a priviledged group, such as CWRU students and astronomers. When the observations are completed, the user is notified and they can download and view the images from the observations.

As of this writing the telescope is publicly available, but some bugs were still being worked out. Exposure times were limited to just a few minutes each, using only the V filter, and some brighter objects, such as the Moon and planets, were off-limits. Case Western has ambitious plans for this telescope, including adding a mosaic camera for wide-field views and a spectrograph to obtain spectra of astronomical objects.

Other Automated Telescopes

The telescopes described in the section above are accessible to anyone on the Internet, without any special requirements or fees. There are other automated telescopes accessible online, but only to specific users. In most cases these are students and teachers, who may have to register, buy specific software, or attend special training sessions in order to gain access to the telescope. We've included these telescopes below in the event that you, or someone you know, is able to use these systems to perform remote astronomy.

MicroObservatory

Harvard University has been developing a network of several small robotic telescopes that, when completed, will be available to students and teachers. The MicroObservatory (**http://mo-www.harvard.edu/ MicroObservatory/**) will have five 6-inch telescopes when completed; as of June 1999 the network consisted of telescopes in Cambridge and Westford, Massachusetts, and Tucson, Arizona, with another being set up at Mauna Kea, Hawaii (Figure 16.7).

The MicroObservatory is an educational effort to introduce students to the night sky and show astronomy is as much a "laboratory science" as chemistry and physics. When completed, students and teachers will be able to enroll in the program and use the telescopes, but while the program is still under development enrollment is by invitation only.

The telescopes will be remotely controlled using a form-based interface on the MicroObservatory Web site. Users submit requests for images, which are prioritized by the system. The images the telescope takes are available for all to view on the Web site, even those not enrolled in the program (Figure 16.8).

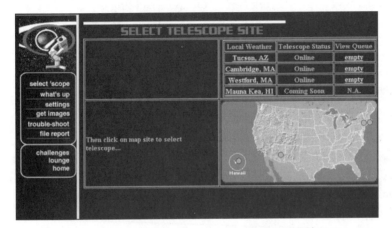

FIGURE 16.7
The MicroObservatory Web site shows the location and status of its current telescopes.

FIGURE 16.8
An image of the M5 star cluster taken with one of the MicroObservatory telescopes

Telescopes in Education

The Telescopes in Education (TIE) project (**http://tie.mtwilson. edu/**) allows students to use a 24-inch telescope at Mount Wilson, California. The project, sponsored by NASA and the Jet Propulsion Laboratory, has been in operation for more than four years. The program is open to any school provided they purchase a copy of the program The Sky: Remote Astronomy Software from Software Bisque to control the telescope.

Unlike other remote telescopes, this system is run by a direct dial-up connection, not over the Internet. This telescope is controlled in real time; instead of submitting a request of an image of an object, you actually use the telescope to point at that object and wait for an image to return. Telescope time can be reserved in blocks from one hour to a whole night. For those unable to observe using the telescope, a gallery of images taken in the past are available (Figure 16.9).

Hands-On Universe

The Hands-On Universe (**http://hou.lbl.gov/telescope/**) program is an educational project at California's Lawrence Berkeley Laboratory that includes, as one component, observations using a

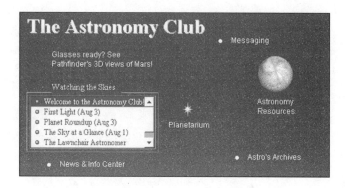

FIGURE 15.1
The Astronomy Club offers a number of resources related to astronomy for America Online members

remote telescope. Students at schools involved in the project can submit observing requests to a 76-cm (30-inch) automated telescope, and later access those images.

Access to the telescope is limited to those schools who have enrolled in the project, which requires teachers to attend workshops to familiarize themselves with the program. The project's Web site includes an archive of images from this and other telescopes that have been involved in the project (Figure 16.10).

Other Telescopes

There are a number of other automated telescopes in use in astronomy, but are not easily accessible by students or the general Internet population. One example is the Katzman Automated Imaging Telescope (**http://astro.berkeley.edu/~bait/kait.html**). This is a 30-inch automated telescope located at Lick Observatory on Mount Hamilton, California. The telescope is used by University of California Berkeley astronomers to hunt for supernovae. It also sees some use in undergraduate astronomy classes at Berkeley (**http://www.ugastro.berkeley.edu/optilab/remote/index.html**), where students can e-mail requests for images for use in class projects. However, the telescope is not open to the general Internet community.

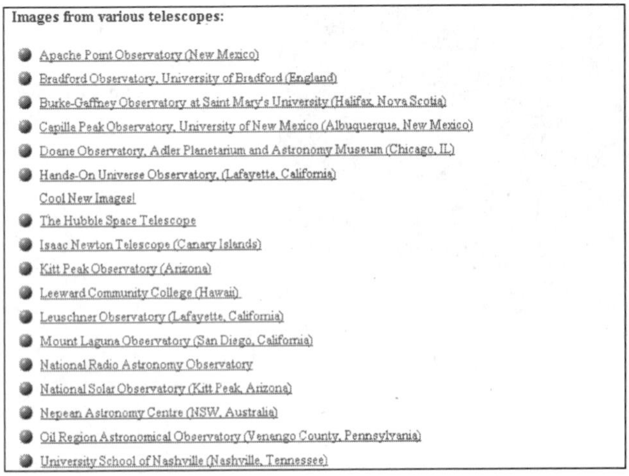

Images from various telescopes:

- Apache Point Observatory (New Mexico)
- Bradford Observatory, University of Bradford (England)
- Burke-Gaffney Observatory at Saint Mary's University (Halifax, Nova Scotia)
- Capilla Peak Observatory, University of New Mexico (Albuquerque, New Mexico)
- Doane Observatory, Adler Planetarium and Astronomy Museum (Chicago, IL)
- Hands-On Universe Observatory, (Lafayette, California)
  Cool New Images!
- The Hubble Space Telescope
- Isaac Newton Telescope (Canary Islands)
- Kitt Peak Observatory (Arizona)
- Leeward Community College (Hawaii)
- Leuschner Observatory (Lafayette, California)
- Mount Laguna Observatory (San Diego, California)
- National Radio Astronomy Observatory
- National Solar Observatory (Kitt Peak, Arizona)
- Nepean Astronomy Centre (NSW, Australia)
- Oil Region Astronomical Observatory (Venango County, Pennsylvania)
- University School of Nashville (Nashville, Tennessee)

FIGURE 16.10
A listing of the telescopes that have provided images in the Hands-On Universe observing program

Similarly, the NF/Observatory at Western New Mexico University (**http://www.nfo.edu/nfo.html**) is a remotely operated telescope used by students and teachers at the university. The Fairborn Observatory (**http://24.1.225.36/fairborn.html**) is a set of remotely operated telescopes in Tennessee provided by several universities. Access to these telescope is limited to the participating universities.

As the technology for running remotely operated telescopes becomes more accessible and less expensive, the number of these facilities will no doubt grow, if for nothing else the convenience of doing astronomy from a Web browser. In the near future students, amateur astronomers, and anyone else on the Internet interested in making their own observations will likely have an even greater number of remotely accessible telescopes on the Internet to choose from!

17

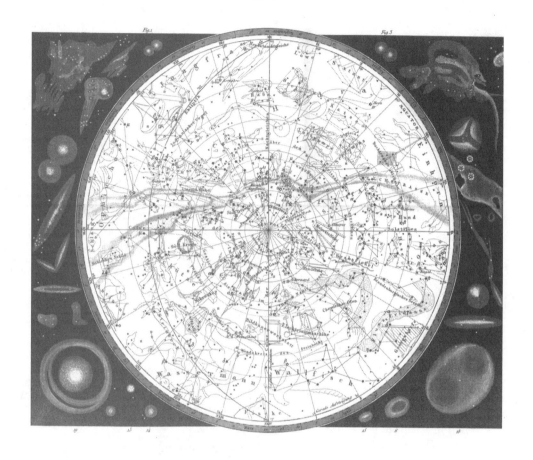

COMPUTERS IN
AMATEUR ASTRONOMY

O bserving the night sky requires no equipment. Simply go out on a clear night and peer into the heavens. However, as your interest in astronomy grows, you'll probably invest in some equipment: binoculars, a telescope, and related accessories. You'll revel in the beauty of the night sky, but you may also feel a bit overwhelmed; how are you going to find the time to look at all the stars, nebulae, planets, and other objects you want to see? You'll soon look for ways to maximize your observing time and look for the best possible objects to observe.

Professional astronomers have a similar urge to maximize their observing efforts. Time at major telescopes is hard to come by, and so astronomers want to make the best use of every minute. Thus, these telescopes are highly computerized, allowing them to quickly locate any object in the night sky. Electronic cameras allow them to quickly take images of their objects and study them almost instantly, giving the observers the opportunity to optimize their observing plans to get as much good data as possible.

Thanks to the growth of personal computers and other electronics, even casual amateur astronomers can study the night sky with a degree of sophistication that would have impressed professional astronomers of just a few decades ago. Computerized telescopes allow astronomers to quickly locate a desired star, galaxy, or other object, without the tedium or frustration of trying to do it manually. Electronic cameras, specifically charge-coupled device (CCD) cameras, allow astronomers to capture images of the night sky on the computer, where they can later analyze and manipulate them, and share them with others on the Internet.

Even before the advent of computerized telescopes and instruments, amateur astronomers have played a major role in astronomy. Since professional astronomers can only observe small regions of the night sky, and because there are far more amateurs than professionals, it's often amateurs who make key discoveries, such as new comets and supernovae. The Internet has made it easier than ever for amateur astronomers to make significant contributions to astronomy in a wide range of fields by making it easy to learn about, and submit observations of, the latest astronomical discoveries.

In this chapter, we'll look at both computerized telescopes and electronic cameras as observing aids for amateur astronomers and at the resources available online for amateurs who want to make scientific contributions to astronomy.

Computer-Controlled Telescopes

Moving a telescope to observe a particular object can often be a frustrating, tedious experience, particularly for beginners. Centering a telescope on a particular star or object can take some time. Finding a dim object may require "star-hopping": moving from one (brighter) star to another to get to the location of the dim object. Moreover, if a telescope is designed to track an object as it moves across the sky, it first must be aligned with the pole star, another time-consuming process.

Major telescopes used by professional astronomers have been computerized for some time. The telescope operator (usually a different person than the visiting astronomer using the telescope) enters the coordinates for a particular object, and the telescope automatically slews to that position. Such automation is required for these telescopes, where every minute of observing time is precious.

This same automation can be applied to amateur telescopes as well. Computerized motors are commercially available that allow users to move the telescope about the night sky by simply entering the coordinates of the object to view. These systems can also be

275

connected to home computers to allow for point-and-click navigation of the heavens.

Several major telescope companies sell telescopes with computerized motors included, or provide them as an add-on option. Meade (**http://www.meade.com**) features the LX200 series of telescopes, ranging in size from 7 to 16 inches (Figure 17.1). These telescopes come with a computerized motor systems that allows them to point towards any desired position in the night sky after the telescope has been aligned by pointing at two known stars. The system also comes with a built-in database of over 60,000 objects that can be accessed by the user.

Meade also produces the Magellan telescope computer system, which allows the user to convert an existing Meade telescope into a computerized system. The Magellan system works similarly to the LX200 system, but with a 12,000-object database.

Celestron (**http://www.celestron.com**) also makes a number of computerized telescopes. The Ultima 2000 is a computerized telescope system with a built-in database of over 10,000 objects (Figure 17.2). The company's Advanced Astro Master system can be used to computerize other telescopes, and has a similar built-in database.

FIGURE 17.1
The 16-inch version of Meade's LX200 computerized telescope

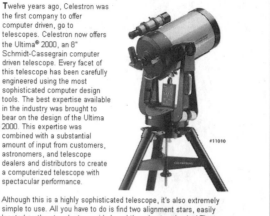

Ultima 11

CG-14

Telescope Warranty

TELESCOPES

PRODUCTS

HOME PAGE

Twelve years ago, Celestron was the first company to offer computer driven, go to telescopes. Celestron now offers the Ultima® 2000, an 8" Schmidt-Cassegrain computer driven telescope. Every facet of this telescope has been carefully engineered using the most sophisticated computer design tools. The best expertise available in the industry was brought to bear on the design of the Ultima 2000. This expertise was combined with a substantial amount of input from customers, astronomers, and telescope dealers and distributors to create a computerized telescope with spectacular performance.

Although this is a highly sophisticated telescope, it's also extremely simple to use. All you have to do is find two alignment stars, easily located on the star charts provided, and the adventure begins! The database contains over 10,000 deep-sky objects and has room for 25 user-defined objects and planets, for easy locating of your favorite

#11010

FIGURE 17.2
Celestron's Ultima 2000 is an 8-inch telescope with a computerized control system

Orion (**http://www.oriontel.com**) sells the SkySensor 2000, an add-on system to computerize existing telescopes. The system works with any GP-mount telescope and includes a database of 6,500 objects. Similarly, the Eureka 2001 system from Maestronix (**http://www.maestronix.com**) can be added to many types of existing telescopes to computerize them. The system includes a database of 15,000 objects.

Software Bisque (**http://www.bisque.com**), a software company, also sells the Paramount GT-1100, a computerized telescope mount (Figure 17.3). Billed as the world's first professional-quality telescope mount designed for consumers, the mount can be used with many existing telescopes to create a computerized telescope system.

You don't need to buy a commercial package to computerize your telescope, if you're willing to work with electronics. *Observatory Techniques* magazine (**http://www.midco.net/~otm**) has a number of articles on building computerized (also known as robotic) telescopes. They also have an extensive set of links to other resources for robotic telescopes at **http://www.midco.net/~otm/ LINKS.HTML**.

Many of these products don't require an additional computer in order to run; the coordinates of the object can be entered into the

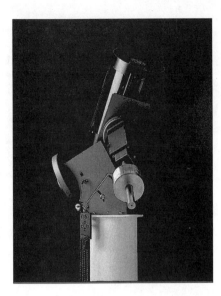

FIGURE 17.3
The Paramount GT-1100 telescope mount, sold by
Software Bisque, can be used to computerize a wide
range of telescopes

telescope's own computer and run directly from there. However,
most provide the option of connecting to your home PC, allowing
you to control your telescope directly from your PC if you have the
right software.

Several major planetarium programs have the ability to operate
telescopes. Software Bisque's The Sky program (Windows 95/NT
and Macintosh) can be used to operate many computerized telescopes, including the LX200 series, the Ultima 2000, and any other
supporting the Astronomical Command Language (ACL) (Figure
17.4). Operating the telescope can then be as simple as selecting an
object in the program and hitting the "Slew To" button.

Sienna Software's Starry Night (**http://www.siennasoft.com**)
also supports a plug-in that allows users to control LX200 series
telescopes and Magellan systems directly from the program. Pointing a telescope at an object is as simple as selecting it in the
program and hitting the "Slew" button (Figure 17.5). Information
on the plug-in, and how to download it, is available at **http://www.
carmelcoast.com/pages/Robin/IG_Astro.html**. (While Starry
Night is available in Windows 95/NT and Macintosh versions, the
plug-in only works on Macs.)

Procyon Systems (**http://www.procyon-sys.com**), developers of the Observer planetarium program for the Macintosh, also make ScopeLink (**http://www.procyon-sys.com/ScopeLink.html**), hardware that can be used to automate a telescope. Such a telescope can then be controlled from within Observer or other PC planetarium programs. The MegaStar planetarium program for PCs (**http://www.willbell.com/software/megastar/index.htm**) can also be used to control LX200 series and compatible telescopes.

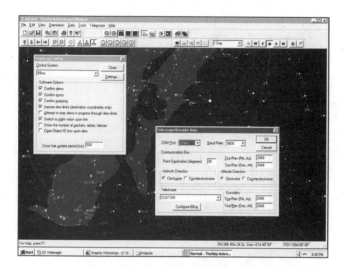

FIGURE 17.4
Software Bisque's The Sky planetarium program can also control computerized telescopes

FIGURE 17.5
A plug-in for Sienna Software's Starry Night program for Macs allows users to control LX200 computerized telescopes

CCD Cameras

With a computerized telescope and a home PC, you could conceivably run your telescope from the comfort of your own home (a very nice option on cold winter nights!), provided you didn't have to go out and actually look through the telescope. However, thanks to the advent of the CCD camera, you can observe the night sky with your telescope and computer without even going outside!

CCDs were first developed at Bell Labs in the late 1960s as memory storage devices. However, researchers found that these devices were better suited as light detectors than memory chips. CCDs were first used in astronomy in the mid-1970s and became commonplace in professional astronomy in the 1980s. In the 1990s the cost of CCDs and related electronics dropped to the point where amateur astronomers could afford their own CCDs camera, some costing as little as a few hundred dollars. See **http:// www.lucent.com/ideas/discoveries/telescope/docs/ccd1.html** for more information on the history of CCDs.

CCDs are semiconductor chips divided up into rectangular grids of pixels that convert light into electrons (Figure 17.6). When a photon of light strikes a CCD, in generates a number of electrons, which are then stored in a "well." These wells are read out when the exposure is completed. The amount of charge accumulated in each well indicates how much light struck that pixel. For a more in-depth introduction to CCDs, check out **http:// star-www.rl.ac.uk/star/docs/sc5.htx/node5.html** and **http://zebu. uoregon.edu/ccd.html**.

Why have CCDs become so widely accepted in astronomy? They have several attributes that make them an almost ideal detector. They are very efficient, capturing up to 80 to 90 percent of the light that falls on them. By comparison, photographic film captures only a few percent of the light it receives, and the human eye even less. Thus, CCDs can take an image of a dim object in far less time than required for a photograph.

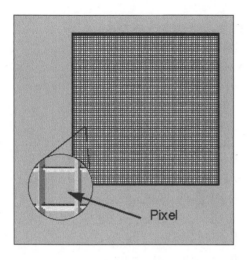

FIGURE 17.6
This illustration, from the Meade Web site, shows how CCDs are composed of a large number of rectangular pixels, which convert photons into electrons

CCD images taken with different filters can be combined to create color images. Since the human eye cannot integrate—collect light over time to build up an image of a dim object—a very large telescope is needed for the eye to be able to see color in anything but the brightest objects. This can be one of the most frustrating experiences for those new to astronomy! However, with a CCD camera and a set of filters, it's possible to create color images using even relatively small telescopes.

CCDs are also linear, meaning an object twice as bright will generate twice as much signal in the CCD. The human eye has a logarithmic response, which allows us to see very bright and very dim objects but makes it difficult to compare the brightness of two objects. CCD images are digital, so they can be manipulated directly by computer. Photos need to be scanned in to be manipulated by computers, and information in the image can be lost in the scanning process.

It's no surprise, then, that CCDs have become commonplace in professional astronomy and are increasingly popular in amateur astronomy. Combining a CCD with a personal computer can make for an instrument with capabilities far beyond what was available in the recent past.

A number of companies sell commercial CCD systems targeted towards amateur astronomers. SBIG, the Santa Barbara Instrument Group (**http://www.sbig.com**) is one of the leading makers of CCD cameras for amateurs and professionals. Their Web site features a great deal of information about their products, as well as software to control the cameras and process the images, a mailing list for their users, and sample images from the CCDs.

Apogee Instruments (**http://www.apogee-ccd.com**) is another leading maker of CCD cameras. In addition to information about their cameras, their site offers links to CCD image analysis software and images taken with their CCDs (Figure 17.7). For those trying to understand some of the intricacies of CCD cameras, the site offers "CCD University," a collection of information on the more technical aspects of CCD cameras. The company is even developing an Internet-accessible remote telescope that will allow prospective customers to control the telescope and take images with Apogee's CCDs. (See Chapter 16 for more information on current Internet-accessible telescopes.)

Starlight Xpress (**http://www.starlite-xpress.co.uk**) is a line of CCD cameras made in Britain designed for amateur astronomers. Its MX5-C CCD camera allows users to take a color image of the night sky in a single exposure, something that usually requires three

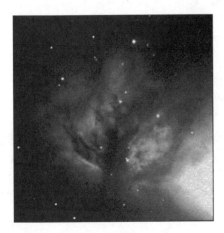

FIGURE 17.7
This image of the Flame nebula, taken by Timothy Puckett, is one of many CCD images available from the Apogee Instruments Web site

exposures taken with different color filters, in other CCDs. Murnaghan Instruments (**http://www.murni.com/astro_0.htm**) makes not only CCD cameras, but a wide array of accessories such as filters, mounts, and other hardware.

Some major telescope makers also sell CCD camera for their telescopes. Meade's Pictor series of CCD cameras (**http://www.meade.com/catalog/pictor**) are designed for amateur astronomers, particularly those using Meade's LX200 series and similar telescopes. Meade also sells *autoguider* CCD cameras, designed to keep the telescope automatically centered on a particular field for long-exposure images by locking on to a bright star visible to the CCD and adjusting the telescope drive to keep the star centered. Celestron's PixCel CCD camera (**http://www.celestron.com/ccd.htm**) uses a CCD from SBIG and includes an internal color wheel to create color images automatically. The Web site includes some sample images from that CCD system (Figure 17.8).

For those in need of more professional CCDs, a number of companies provide high-quality CCDs—at a price. Photometrics (**http://www.photomet.com**) developed the first commercial CCD camera near 20 years ago and is a leading maker of high-performance CCDs today. Its Web site includes not only information about its

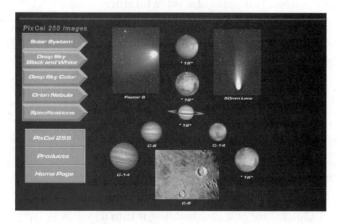

FIGURE 17.8
Celestron's Web site includes CCD images of planets and other objects in the night sky, taken with their CCD cameras

instruments but images and a reference library of information about the technical aspects of CCDs.

Scientific Imaging Technologies (**http://www.site-inc.com**) produces high-grade CCDs, and its Web site provides information on these products as well as a tutorial on how CCDs work. A similar company, PixelVision (**http://www.pv-inc.com**), also has technical papers about CCD operations as well as a description of its products. EEV Ltd. (**http://www.ccd.eev.com**), Thomson CSF (**http://www.tcs.thomson-csf.com/Us/ccd/ccd.htm**), and Andor Technology (**http://www.andor-tech.com**) also make high-end CCDs, largely for professional use.

Those interested in the development and use of CCDs by professional astronomers will also be interested in CCD-world, a mailing list about CCD development. The list has a Web site at **http://www.not.iac.es/CCD-world/** that includes mailing list information, archives of past messages, and links to other CCD resources worldwide.

While CCD camera can be an expensive proposition, they can also be made quite cheaply. Richard Berry's *CCD Cookbook* is a popular book that describes how to build a simple CCD camera that's more than sufficient for most amateur astronomers, for only a few hundred dollars. The CCD Cookbook Web site, at **http://wvi.com/~rberry/cookbook.htm**, provides some information about the project and answers some commonly-asked questions about such a CCD camera, with links to other Web resources, including companies like University Optics (**http://www.universityoptics.com**), who sell ready-made parts for the CCD Cookbook camera.

Amateur astronomy CCD cameras can also be adapted from the growing market in consumer digital cameras. Many people have successfully built a CCD camera using a Connectix QuickCam camera (**http://www.connectix.com**), which is available in a grayscale version for less than $100. This process requires some work disassembling the original QuickCam and reassembling it into a mount suitable for a telescope. This procedure is described at

http://perso.club-internet.fr/uranos/disassemble_quickcam.htm,
with other information, and images taken with such a system, at
http://www.geology.ewu.edu/jpb/lho/lho.htm (Figure 17.9).

Once you've set up a CCD camera to take images, you'll want
some way to view, manipulate, and store the images for future use.
A number of software packages exist to do this. Software Bisque's
CCDSoft (http://www.bisque.com) allows users to do a wide
range of analysis on CCD images, as well as operate the CCD camera itself (Figure 17.10). It can also work with the company's The
Sky program to identify objects visible in the CCD images.

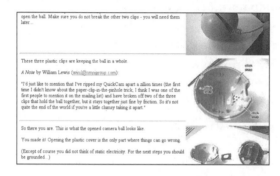

FIGURE 17.9
Web sites provide information on how to
adapt a QuickCam digital camera into a
CCD camera suitable for amateur astronomy

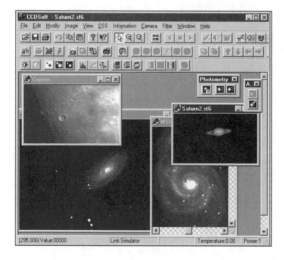

FIGURE 17.10
CCDSoft is a software package that allows
users to perform sophisticated analyses on
their CCD images

If your computer runs a version of the Unix operating system, check the Clear Sky Institute (**http://www.clearskyinstitute.com**). They develop astronomy image analysis software for Unix machines, including OCAAS, the Observatory Control and Astronomical Analysis System, a complete software package that can operate remote telescopes, run CCD cameras, and analyze images taken by the system.

Mira, from Axiom Research (**http://www.axres.com/~axiom/**), is an image processing program for PCs that comes in several versions, from a lightweight version ideal for amateurs to advanced versions for professionals. MAIA, a shareware astronomical image analysis package for Macintosh computers, can be downloaded from **http://www.deepspace.ucsb.edu/util/** along with related software for PCs and Macs. A wider list of software packages for various computer systems is available at **http://www.isc.tamu.edu/~astro/software.html**.

If you need some heavy-duty image analysis, such as the type professional astronomers use, look into IRAF (**http://iraf.noao.edu/**), the Image Reduction and Analysis Facility. This software, developed at the National Optical Astronomy Observatory, has the ability to do all sorts of sophisticated analysis on CCD images, certainly far more than what a typical amateur astronomer would find interesting. With this power, however, comes a steep learning curve; IRAF is not easy to use, for beginners and even some experienced users. It's available for PCs and various Unix systems.

An Alternative to IRAF is IDL, the Interactive Data Language, from Research Systems, Inc. (**http://www.rsinc.com/idl/main.html**). This is a general-purpose data analysis package, for which a number of astronomy-specific routines have been developed; a library of these routines are available at **http://idlastro.gsfc.nasa.gov/homepage.html**. However, IDL is not cheap; a full license can run into the thousands of dollars.

Scientific Resources for Amateurs

Many amateur astronomers are quite content to explore the night sky on their own, observing whatever strikes their fancy at the time. Others, though, are interested in expanding the frontiers of knowledge. Astronomy is one of the few remaining fields of science where amateurs can and often do make significant contributions, from the monitoring of variable stars to the discovery of new comets and supernovae. Amateurs have been making such contributions for many years, but thanks to the Internet it is easier than ever for interested observers to keep up to date with the latest discoveries and plan their observations accordingly, or to communicate possible new discoveries with the appropriate officials. Some of the best resources online in these fields are discussed below.

Solar, Lunar and Planetary Observations

The solar system is a dynamic place, in need of constant observation to keep up with all its changes. These changes range from the movement of sunspots on the Sun to the merging of storms on Jupiter and the formation of dust storms on Mars. It's not always possible for professional astronomers to keep track of all this action at once, so they rely on the amateur astronomy community to observe and report on these changes.

One group through which much of this work is done is the Association of Lunar & Planetary Observers (ALPO), a group of about 600 astronomers located worldwide. These astronomers specialize in all aspects of Solar System observations, from views of the Sun and Moon to planets, comets, and asteroids.

Their Web site, at **http://www.lpl.arizona.edu/~rhill/alpo/ index.html**, has membership information as well as news and information from each of its "sections," subgroups that focus on particular Solar System objects. The pages for each section include news about recent observations, report forms, and related resources. There is also information on ALPO's training program, an effort to train amateur astronomers to conduct productive and meaningful scientific observations.

Asteroids and Comets

Of particular interest to Solar System astronomers is the study of asteroids and comets. While much of the focus of this work is on the discovery of new objects, amateurs plan a critical role by observing existing objects to refine their orbits. Both the discovery and follow-up of these objects are key areas of contributions by amateur astronomers.

The key resource for this work is the International Astronomical Union's Minor Planet Center (MPC), located at the Smithsonian Astrophysical Observatory and on the Web at **http://cfa-www.harvard.edu/iau/mpc.html**. The MPC collects observations of asteroids and comets, computes orbits based on these observations, and disseminates the results to the astronomical community. The Web site has information about how to submit astrometric observations—the positions of asteroids and comets—to the MPC for their analysis.

Discoveries of new asteroids should be reported to the MPC, whose Web site includes contact information and a search tool to check for known asteroids in a specified region of the sky, to be sure you have not rediscovered an existing asteroid! Discoveries of comets, however, are made through a related office, the Central Bureau of Astronomical Telegrams (**http://cfa-www.harvard.edu/iau/cbat.html**). The site has detailed information regarding how to submit announcements of possible comet and other discoveries, including an online form for submitting discoveries (Figure 17.11). The site also has information on the Edgar Wilson Award, a new cash prize for rewarding comet discoveries by amateur astronomers.

Occultations

A specialized class of observations particularly suited to amateur astronomers is occultations, when one body—a planet, moon, or asteroid, generally—passes in front of a more distant star as seen from the Earth, cutting off the light from the star for a brief time. By timing the exact beginning and end of these occultations at multiple locations, it's possible to determine the shape and size of

Form For Reporting Items To The Central Bureau For Astronomical Telegrams

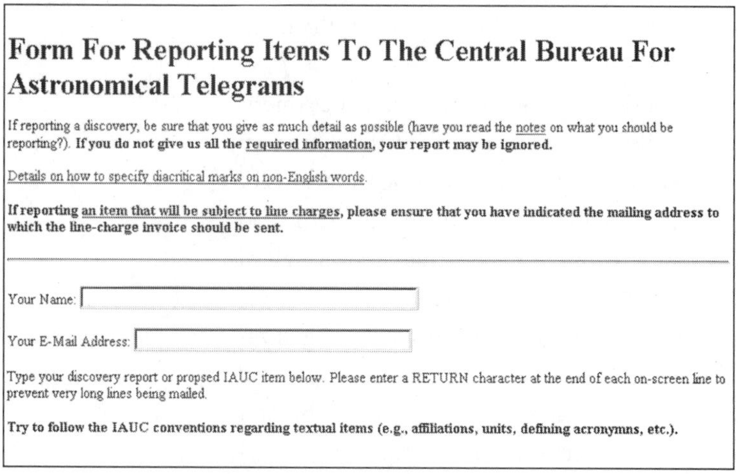

If reporting a discovery, be sure that you give as much detail as possible (have you read the notes on what you should be reporting?). **If you do not give us all the required information, your report may be ignored.**

Details on how to specify diacritical marks on non-English words.

If reporting an item that will be subject to line charges, please ensure that you have indicated the mailing address to which the line-charge invoice should be sent.

Your Name:

Your E-Mail Address:

Type your discovery report or propsed IAUC item below. Please enter a RETURN character at the end of each on-screen line to prevent very long lines being mailed.

Try to follow the IAUC conventions regarding textual items (e.g., affiliations, units, defining acronyms, etc.).

FIGURE 17.11
Reporting a key astronomical discovery can easily be done with this form on the CBAT Web site

the occulting body, a particularly powerful way to study asteroids. Grazing lunar occultations, where the edge of the Moon just passes in front of a star, can be used to refine the positions of the star and Moon as well as study the lunar topography, as the star disappears and reappears among the mountains and valleys along the visible edge of the Moon.

The key source of information for amateur astronomers interested in the subject is the International Occultation Timing Association (IOTA), on the Web at **http://www.occultations.org**. The Web site has information about joining IOTA as well as links to its two main sections, the asteroid occultation section (**http://www.anomalies.com/iotaweb**) and the lunar occultation section (**http://www.lunar-occultations.com/iota/iotandx.htm**). Each site has information on past and future occultations, observing information, and how to report your occultation observations.

Variable Stars and Supernovae

If it's difficult for the professional astronomical community to keep with the relative handful of objects in the solar system, you can imagine how hopeless it is for them to keep up with the many more stars in our galaxy. Amateur astronomers play a key role here doing not only routine observations of stars but also keeping on the lookout for sudden changes in these stars.

The primary organization that supports this work is the American Association of Variable Star Observers (AAVSO), located on the Web at **http://www.aavso.org**. It coordinates observations of variable stars from several hundred astronomers, mostly dedicated amateurs, worldwide (despite the organization's name, over half of its observers are located outside the United States.) It receives over 350,000 observations a year, and over 9 million since the organization was founded in 1911. The AAVSO site has background information about variable stars, the latest news about variable star observations, online data, and information about reporting observations to the AAVSO (Figure 17.12).

Amateurs also play a key role in the discovery of supernovae. Able to patrol much larger regions of the sky than professionals, amateurs are often the first to discover new supernovae. The International Supernovae Network (**http://www.supernovae. net-isn.htm**) has information about the latest supernova discoveries and a detailed set of guidelines for observers to follow if they think they've discovered a supernova. Reports of supernova discoveries, like those for comet discoveries, should go to the IAU's Central Bureau of Astronomical Telegrams. Their Web site

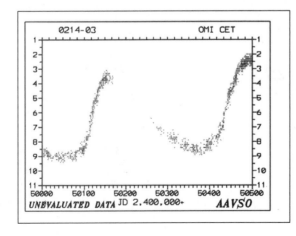

FIGURE 17.12
The AAVSO Web site includes data on variable star observations, including this light curve (brightness versus time) of the variable star Mira

(http://cfa-www.harvard.edu/iau/cbat.html) has information regarding how to report a discovery, including tools for checking a possible supernova discovery against a database of known asteroids.

As you can see, computers and the Internet can play a vital role for amateur astronomers, whether your telescope is completely computerized or if you simply use computers to check on and report astronomical discoveries. By combining a computerized telescope with a CCD camera, you can create a system that allows one to view the night sky without being out underneath the night sky. While the observer might just be a short distance away, in the warm comfort of the indoors, the observer could also be hundreds or thousands of miles away, connecting remotely by modem or over the Internet.

APPENDIX A

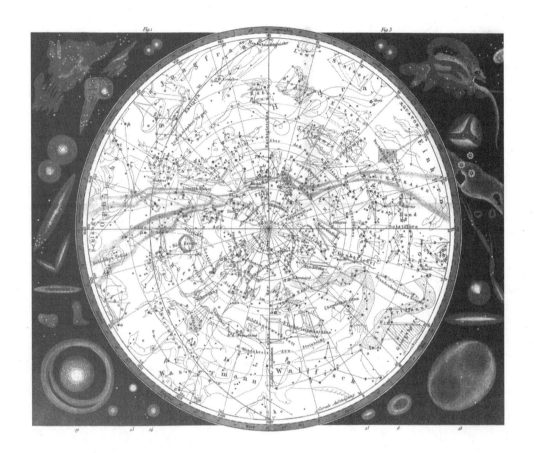

SOFTWARE ON THE
CD-ROM

The CD-ROM contains an extensive collection of astronomy-related software. This appendix describes each product on the CD and also gives directions on how to install the programs. Programs marked *freeware* are free of charge; for those marked *shareware*, you may use the program for free on a trial basis, but then you must pay to register the software. In the case of shareware programs, read the documentation for each program you want to register to find out how much it costs.

Note: In general, products marked as DOS products will run under Windows, but we cannot guarantee the operation of any program. We've included instructions for running DOS programs from a DOS prompt (perhaps the safest option), but in most cases, you should encounter no problems running DOS programs directly from within Windows. To do so, click the named file in the instructions for each installation—typically an .exe or .com file.

Installation Notes

DOS

We recommend that you copy DOS programs to a new directory on your hard disk, and that you install from there. We do not recommend installing DOS programs from the CD-ROM, as such installation attempts will most likely fail. The instructions here assume that for each DOS program you have copied the necessary files to your hard disk, and that you are installing from there.

Windows

We assume that you are probably running Windows 95 or higher. If so, you should be able to install most Windows programs directly from the CD-ROM, according to the installation instructions given here. We do not recommend running programs from the CD-ROM.

Macintosh

Installation of Macintosh programs is usually quite simple, as reflected in their instructions here.

Unzipping Files

If you don't have a decompression program installed on your machine (like Winzip), try the free decompression programs from Aladdin Systems: Stuffit Expander (Windows 3.1) or Aladdin Expander (Windows 95 or higher). You'll find them below, listed under Utility Programs. When you install them, we suggest that you accept the default options. Once installed, double-clicking on zipped or otherwise compressed programs should launch StuffIt or Aladdin to unpack/uncompress the files within them.

To create a new directory (folder) on your hard disk

DOS

From a DOS prompt type **MD <newdirectoryname>** and press **ENTER.**

Windows 3.1

Double-click the File Manager icon and then select "Create Directory" from the File menu.

Windows 95

Open Windows Explorer and choose File • New • Folder. When a new folder with a blank name tag appears enter a name.

What If I Have Trouble?

Should you encounter problems running any of the software on the CD-ROM, please contact the person or company listed in the individual program descriptions.

Note: We assume throughout that you know how to create new directories on your hard disk and copy and unzip files into them. We also assume that you know how to run programs from a DOS prompt, and that you are comfortable running Windows or DOS setup or installation files.

PC Software

Atmospheric Conditions Programs

ATMOS (. ./PC/ATMOS)
Contact: David Eagle
Type: Freeware
Operating System: DOS

Computes a typical atmospheric profile for the Earth. Enter the starting and ending heights, in meters, and a distance increment; then, for each increment, the program computes and displays the conditions at each level for a standard model atmosphere, showing the pressure, density, and temperature at each level.

1. Create a new directory on your hard disk called **ATMOS**, copy **atmosph.com** into it, then run **atmosph.com**.

2. Exit to DOS, change to the **ATMOS** directory, and enter **ATMOS** to run the program.

WetSock v. 3.8b (. ./PC/WETSOCK)

Contact: Locutus Codeware; Email: info@locutuscodeware.com; Web: http://www.locutuscodeware.com
Type: Shareware
Operating System: Windows 95 or higher

WetSock brings weather information to your desktop. Click on the WetSock icon in your taskbar to see your the local weather conditions and forecasts for almost 3,000 cities worldwide.

1. Run **wetsocks.exe** to install.

*Note: This version of WetSock requires the updated Windows Common Controls .dll from Microsoft. If you do not already have this installed on your Windows 9.x or Windows NT system, run **40comupd.exe** found in the . . PC/40comupd/ directory on this CD-ROM and follow the onscreen directions.*

WinWeather v. 4.0 (. ./PC/WINWEATH)

Contact: Insanely Great Software, Attn: Adam Stein, 126 Calvert Ave. E., Edison, NJ, 08820; Phone (support): (617) 266-1630; Sales (orders): (800) 319-5107; Fax: (732) 632-1766; Web: http://www.igsnet.com/weather.html
Type: Freeware or $19.95 for the Pro version (shareware)
Operating System: Windows 3.1 (weath16.exe) or Windows 95 or higher (weath32.exe)

Covers a broad range of weather needs. Get live pictures from both American and international weather cameras, hourly weather reports and forecasts, the latest satellite images, and such exotic items as the ultraviolet light forecast. An excellent application that can track an unlimited list of cities, saving the data for later review and evaluation.

1. Run **weath16.exe** (for Windows 3.*x* systems) or **weath32.exe** (for Windows 9.x/NT systems).

Clocks

ASTROCLK v. 9848 (. ./PC/ASTROCLK)

Contact: David H. Ransom, Jr., 240 Bristlecone Pines Road, Sedona, Arizona 86336; Email: rans7500@spacelink.nasa.gov; Web: http://www.dransom.com
Type: Shareware
Operating System: DOS

This astronomical clock and celestial tracking program shows the current time in a wide variety of systems such as local, UTC/GMT, solar, sidereal, and GPS. ASTROCLK also tracks the stars and planets, and users can perform dead reckoning navigation and reduce star sights. Updated from the *Astronomical Almanac* annually.

1. Create a new directory on your hard disk called ASTROCLK and copy all files into it.

2. Run **astroclk.exe**. (See **win95.txt** for instructions on configuring ASTROCLK for use under Windows 95.)

Astronomy Clock 2 (. ./PC/ASTRCLK2)

Contact: Personal MicroCosms/Pocket-Sized Software, 8547 E. Arapahoe Road, Suite J-147, Greenwood Village, CO 80112; Email: EricTerrell@hotmail.com; Web: http://www.ericterrell.simplenet.com/html/ss_ac32.html
Type: Shareware
Operating System: Windows 95 or higher

A simple but useful time utility that displays the current time in local mean time, Universal Time, local sidereal time, and Greenwich sidereal time.

1. Run **setup.exe** to install.

GeoClock (. . PC/GEOCLOCK)

Contact: Joseph R. Ahlgren; Email: Joe@GeoClock.com; Web: http://www.clark.net/pub/bblake/geoclock
Type: Shareware
Operating System: DOS (geoclk82.zip, version 8.2); Windows 3.1 or higher and OS/2 (Geoclk81.zip, version 8.1)

Uses the computer's current time to generate a world map, showing which parts are in daylight. Also displays times around the world as well as sunrise and sunset times, Moon phases and positions, and Moon rising and setting times. A quick and easy way to understand the relative position of both the Sun and Moon in relation to the Earth and the seasonal differences in day length due to the Earth's axial tilt. Installation instructions and extensive documentation are included in each of the program archives. Additional maps are available in **geoxtr82.zip**, and the most current time zone file available when we completed this CD-ROM is included in **geozones.zip**.

DOS:

1. Create a directory on your hard disk called GEOCLK and unpack all the files in **geoclk82.zip**, **geoxtr82.zip**, and **gckwin82.zip** into it.

2. Exit to DOS, change to the GEOCLK subdirectory, and enter **GEOSETUP** to customize the program for your location and hardware. **GEOCLK** runs the program.

Windows

1. Create a directory on your hard disk called GEOCLK and unpack all the files in **gckwin81.zip**, **geoxtr82.zip**, and **geoclk82. zip** into it.

2. Run **GEOCKWIN** and choose **File • Setup** to customize the program for your location and hardware.

SocketWatch v. 3.2 (. . PC/SKTWATCH)
Contact: Locutus Codeware; Email: info@locutuscodeware.com; Web: http://www.locutuscodeware.com
Type: Shareware
Operating System: Windows 95 or higher

Runs in the background and automatically synchronizes your PC clock with atomic clocks.

1. Run **sswatch.exe** and follow the onscreen directions.

SUNDEMO v. 1/97 (. . PC/SUNDEMO)
Contact: Gianni Ferrari, Via Valdrighi 135, 41100 Modena, Italy; Email: frank.f@pianeta.it
Type: Shareware
Operating System: DOS

Simulates sundials including the classical as well as monofilar and bifilar versions and those on a cylinder or cone.

1. Create a directory called **SUNDEMO**; then unzip **sundemo.zip** into it.

2. Exit to DOS, change to the **SUNDEMO** subdirectory, and type **SUNDEMO** at the DOS prompt. Read **sundemo2.txt** for additional instructions.

Flight Simulators

Basics of Space Flight (. . PC/BSFDEM)

Contact: Leo P. Gaten, c/o Information Services, P.O. Box 811, Sequim, WA 98382; Email: lgaten@olympus.net; Web: http://www.olympus.net/personal/lgaten/index.htm
Type: Demoware
Operating System: Windows 3.1 or higher

A book-style introduction to the science and engineering aspects of space flight, based on information provided by JPL (Jet Propulsion Laboratories). The sample chapters provide basic information about the Solar System, gravitation, orbits and trajectories, and electromagnetic phenomena.

1. Create a directory or your hard disk called **BSFDEM** and unzip **bsfdem10.zip** into it.

2. Run **install.exe** to install the software.

Space Flight Simulator (. . PC/FLGHTSIM)

Type: Freeware
Operating System: DOS

Simulates space flight in the inner Solar System. The system displays your spacecraft relative to various planets in the inner Solar System. Enter a direction and acceleration to change course. A small change in acceleration can have a big effect on your trajectory!

1. Create a directory on your hard disk called **SPACEFLT** and copy **spaceXt.com** into it.

2. Exit to DOS, change to the **SPACEFLT** subdirectory, and type **SPACEFLT**.

Galaxy Programs

CLEA Exercise—Hubble Redshift v. 0.60 (. . PC/CLEAHUB)

Contact: Project CLEA; Phone: (717) 337-6028; Email: clea@ gettysburg.edu
Type: Freeware
Operating System: Windows 3.1 or higher

An educational program that simulates the operation of a telescope studying galaxies. Select a field to observe, set up the telescope, take spectra of selected galaxies, and then compare the wavelengths of features in the spectra to measure redshift. Use these redshift measurements, together with the apparent and absolute magnitudes of the galaxy provided in the program, to determine the Hubble constant.

1. Create a directory called CLEA on your hard disk, unzip **clea_hub.zip** into it, and then run the program **Clea_hub.exe**.

GALACTIC SLAM DANCE (. . PC/SLMDANCE)

Contact: Douglas E. Music; Email: music@indirect.com
Type: Freeware
Operating System: DOS

Simulates the effects of galactic collisions by tracking the scattering of stars. Specify the number of stars in the target galaxy, the ratio of the masses of the target and intruder galaxies, and the original location of the intruder galaxy and the velocity. The program then calculates the motion of the intruder galaxy relative to the target, modeling the effects of the intruder galaxy's gravity on the scattering of stars and graphing the results.

1. Create a directory called **GC3D** on your hard disk and unzip **gc3d.zip** into it.

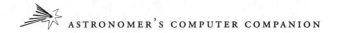

2. Exit to DOS, change to the **GC3D** directory, and then type **GC3D** to run the program.

Galaxy Collisions (. . PC/COLLIDE)
Contact: Jeff Alderman
Type: Freeware
Operating System: DOS

Simulates the collision of two galaxies. You determine the number of stars in the galaxy and the mass and velocity of the intruder galaxy.

1. Create a directory called **COLLIDE** on your hard disk and copy all files into it.

2. Exit to DOS, change to the **COLLIDE** subdirectory, and enter **GALAXY** to run the program. See **read.me** for more information.

Three-Dimensional Milky Way Viewer (. . PC/3DMW)
Contact: Rick Arendt; Web: http://home/pacbell.net/wolton
Type: Freeware
Operating System: DOS

A 3D fly-through, user-controlled model of the Milky Way. Shows up to 5,000 stars (with each star representing up to 100 million actual stars). Run it in demo mode or use the mouse to change the direction and viewing angle.

1. Create a directory on your hard disk called **3DMW** and copy all files into it.

2. Exit to DOS, change to the **3DMW** subdirectory, and type **3DMW** to run the program.

Utility Programs

Adobe Acrobat Reader (. . PC/ADOBE)
Contact: Adobe Systems; Web: http:\\www.adobe.com
Type: Freeware
Operating System: Windows 3.1 (ar16e301.exe);Windows 95 or higher (ar40eng.exe)

Use this program to view Portable Document Format (PDF) files contained on this CD-ROM.

1. Run a **ar16e301.exe** to install the Windows 3.1 version; run **ar40eng.exe** for Windows 95 or higher and NT.

Aladdin Expander 5.0 (. . PC/Stuffit/32-bit)
Contact: Aladdin Systems; Web: http://www.aladdinsys.com
Type: Freeware
Operating System: Windows 95 or higher

A Windows decompression tool that gives access to all StuffIt and Zip compressed files. Offers 32-bit, long file name support and is Windows 95/98/NT 4.0 compatible.

1. Run **alex50.exe** to install.

Microsoft Common Control .DLL update (. . PC/Mcc)
Contact: Microsoft, http://www.microsoft.com
Type: Freeware
Operating System: Windows 95 or higher

This update from Microsoft is required by certain applications on this CD-ROM.

1. Run **40comupd.exe** to install.

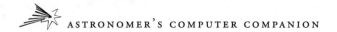

QuickTime 4 (. .PC/QTIME)

Contact: Apple Computer, Inc., Web: http://www.apple.com
Type: Freeware
Operating System: Windows 95 or higher. Windows 3.1 customers should continue to use QuickTime 2.1.2.

QuickTime 4 is freeware, but registering the application turns it into the Pro version and enables many extra features.

QuickTime is an Apple technology that lets you view and create video, sound, music, 3D, and virtual reality for both Macintosh and Windows systems. Several of the animations and panoramas found on this CD-ROM require QuickTime for viewing.

1. Run **QuickTimeInstaller.exe** to install.

StuffIt Expander 1.0 (. . PC/Stuffit/16-bit)

Contact: Aladdin Systems; Web: http://www.aladdinsys.com
Type: Freeware
Operating System: Windows 3.1

A Windows decompression tool that gives access to all StuffIt and Zip compressed files. It provides quick and easy file access, regardless of the platform it was created on, its origin, or how it was transmitted.

1. Run **sitex10.exe** to install.

Gravity Programs

Gravitational N-Body Simulation (. . PC/N-BODY)

Contact: Henley Quadling; Email: HQUADLING@mnhep.hep.umn.edu
Type: Freeware
Operating System: DOS

Simulates "N-body dynamics"—the way a set of bodies interacts with one another gravitationally. Create particles, define their masses and initial velocities and positions, and control the time step and accuracy of the calculations.

1. Create a directory on your hard disk called **N-BODY** and copy all files into it.

2. Exit to DOS, change to the **N-BODY** subdirectory, and enter **NBODYD16** to run the program.

Gravity v. 1.0 (. . PC/GRAVITY1)

Contact: George Moromisato, TMA, 15 Whittier Rd., Natick, MA 01760
Type: Freeware
Operating System: Windows 3.1 or higher

This gravity simulator allows you to set up a number of different bodies with different initial masses, velocities, and positions to see how the bodies interact with each other. Create new bodies by clicking where the body should start; then set their velocity and direction by clicking and dragging the mouse. Includes sample simulations, including simulations of our inner Solar System and a binary star system.

1. Create a directory called **GRAVITY1**, copy **gravity1.exe** into it, and then run **gravity1.exe** to unpack the software.

2. Run **gravity.exe** to begin the program.

GRAVITY v. 1.2 (. . PC/GRAVITY2)

Contact: Jens Jorgen Nielsen, Studsgaardsgade 48, DK-2100 Copenhagen East, Denmark; Email: jjn@login.dknet.dk
Type: Freeware
Operating System: DOS

Simulates the gravitational effects on a set of bodies in three dimensions. Define the mass, initial position, and velocity of each object as well as other parameters like the time step, viewing mode, and color of each object. Includes a number of predefined parameter files for modeling the orbit of the Moon, a solar system, and a galaxy collision.

1. Create a directory on your hard disk called **GRAVITY2** and copy all files into it.

2. Exit to DOS, change to the **GRAVITY2** subdirectory, and enter **GRAVITY** to run the program. Documentation is included in the **gravity.txt** file.

Gravity Simulator v. 1.0 (. . PC/GRAVSIM)

Contact: Kevin Cheung; Email: cheungi@mskcc.org
Type: Shareware
Operating System: Windows 3.1 or higher

Simulates the orbital path of a satellite on an almost immovable mass and allows you to change several parameters, including the masses of the Earth and satellite and the satellite's initial position and velocity. Plots the satellite's orbit around the Earth, including the vector of the effective gravitational force on the satellite and the satellite's velocity vector. Adjust the program's number of computations to change the speed of the simulation. **Note:** Requires Visual Basic 4.0 runtime libraries (see . . PC/VBRUN400/VB40032A.ZIP).

1. Create a directory on your hard disk called **GRAVSIM** and copy all files into it

2. Run **grav16.exe** to start the program.

Newton v. 1.0 (. . PC/NEWTON)

Contact: Paul Keet, 12726 Southridge Dr., Surrey, B.C., V3X 3C6, Canada
Type: Shareware
Operating System: Windows 3.1 or higher

This gravity simulator lets you create a set of bodies with masses, positions, and velocities and then run simulations to see how the bodies interact. (Control the time steps to speed up or slow down the program.) Will also plot the trails of moving objects, display grid positions, and show velocity vectors. Includes some sample simulations, including the Moon orbiting the Earth and the effects of intruder bodies on a stable planetary system.

1. Create a directory on your hard disk called **NEWTON**, copy all files into it, then run **Newton.exe**.

TIDES v. 3.09 (. . PC/TIDES)

Contact: Nautical Shareware, Edward P. Wallner, 32 Barney Hill Road, Wayland, MA 01778-3602; Phone: (508) 358-7938; Email: epwallnr@world.std.com
Type: Shareware
Operating System: DOS

Computes estimated high- and low-tide levels and times for numerous locations. Will also display the astronomical conditions that contribute to tides, including the location of the Sun and Moon.

1. Create a directory on your hard disk called **TIDES** and then unzip **tides309.zip** into it.

2. Exit to DOS, change to the **TIDES** subdirectory, and type **TIDES** to run the program. Read **tides.doc** for details on the program and its operation.

Image Analysis Programs

Astroart Demo Version (. . PC/ASTROART)

Contact: MSB di F.Cavicchio, Via Romea Vecchia 6748100 Classe (RA), Italy; Fax: ++39 0544-473589; Email: msb@ntt.it; Web: http://www.sira.it/msb/astroart.htm

Type: Demo

Operating System: Windows 95 or higher

Analyzes and processes astronomical images. Includes a built-in star atlas based on the G.S.C. catalog to perform astrometric and photometric calibrations of images. Supports both the 8- and 16-bit FITS file formats produced by commercial CCD cameras and can import generic file formats up to 4000 x 4000 pixels. A wide range of pre- and post-processing filters let you get the most from each image. Can also be used as a blink comparitor.

    1. Create a directory called **ASTROART** and unzip **aadem_en.zip** into it; run **astroart.exe** to use the demo.

UCFImage v. 5.00 (. . PC/UCFIMAGE)

Contact: Harley Myler, Department of Electrical and Computer Engineering, University of Central Florida, Orlando, FL 32816

Type: Freeware

Operating System: DOS

This basic image analysis program can read or write images in BMP, GIF, PCX, and TIF formats and can perform several operations on those images, including Fourier transforms and various types of filtering. It will also measure the images and display histograms.

    1. Create a directory on your hard disk called **UCFIMAGE** and copy the contents of the **UCFIMAGE** directory to it.

2. Exit to DOS, change to the **UCFIMAGE** directory, and enter **install** to install the program.

3. Run the program by typing **run**. See **read.me** for more information.

Multipurpose Programs

Deepsky 99 v. 2.1.8 (. . PC/DPSKY99)

Contact: Steven S. Tuma, 1425 Greenwich Ln., Janesville, WI 53545; Phone: (608) 752-8366; Email: stuma@inwave.com; Web: http://www.deepsky2000.com

Type: Demo
Operating System: Windows 95 or higher

This terrific program offers a planetarium, an observing log, telescope control, some astronomical image processing capabilities, and much more. This demo version includes the 13,000-object database (the Hubble Guide Star Catalog of stars to 15.5 magnitude is available on CD-ROM only), but does not include the massive database of images—which is available for download at the Web site.

1. To install, run **ds99073098.exe** and follow the directions on your screen. Documentation is contained in a Microsoft word document included in **userinformation.zip**.

Deep Space v. 5.56 (. . PC/DPSPACE)

Contact: David Chandler Company, P.O. Box 999, Springville, CA 93265; Phone: (559) 539-0900; Fax: (559) 539-7033; Email: david@davidchandler.com; Web: http://www.davidchandler.com

Type: Demo
Operating System: DOS

Deep Space is an outstanding observing guide that generates publication-quality star maps; serves as an almanac, ephemeris, and observation log; controls telescopes; and more. One of the best for DOS.

1. Copy all files into a temporary directory on your hard disk and run **install.exe**.

SkyChart III v. 3.11 (. . PC/SKYCHART)
Contact: Southern Stars Software, 12525 Saratoga Creek Drive, Saratoga, CA 95070; Phone: (408) 973-1016; Fax: (408) 973-1016; Email: info@southernstars.com ; Web: http://www.southernstars.com
Type: Demo
Operating System: Windows 3.1 or higher. The demo version included here is only for Windows 95 or higher.

This excellent and reasonably priced Astronomy application includes a wealth of features and is available for a wider variety of computing platforms than similar programs. In addition to providing an excellent planetarium function, SkyChart III can compute and display the positions of Earth satellites using standard NASA/NORAD two-line element satellite orbit files and print high-resolution star charts and highly detailed ephemerides. It also includes a fully customizable object database containing more than 40,000 stars and hundreds of Messier, NGC, and IC deep-sky objects. All existing objects can be edited, including the constellations, and you can import your own database to add new comets, asteroids, stars, planets, or deep-sky objects.

1. To install the demo version, run **SCD311.exe**.

Movie Makers

Astronomy Lab 2 v. 2.03 (. . PC/ASTROLAB)
Contact: Personal MicroCosms/Pocket-Sized Software, 8547 E. Arapahoe Road, Suite J-147, Greenwood Village, CO 80112; Email: EricTerrell@hotmail.com; Web: http://www.ericterrell.simplenet.com/index.html
Type: Shareware
Operating System: Windows 95 or higher

Generates animated movies that simulate a host of astronomical events (including solar and lunar eclipses, lunar occultations, planetary occultations, transits of Mercury and Venus, the orbits of Jupiter's bright moons, and the motions of the planets in the plane of the ecliptic). Also produces several reports on the most important and exciting astronomical events of the past, present, and future. A graphing feature illustrates many fundamental astronomical concepts.

1. Run **setup.exe** to install the program.

Observational Programs

AA v. 5.4 (. . PC/AA)
Contact: Stephen Moshier; Email: moshier@world.std.com
Type: Freeware
Operating System: DOS

Computes and displays detailed information about planets, stars, and other bodies. Enter the time and the particular object, and the number of times the information should be computed to display detailed information from the object's ephemeris, including its tight ascension and declination, distance from the Earth, diunral parallax and aberrations, and more. Includes a built-in ephemeris of all nine planets as well as the Sun and can also import information from separate data files. Includes a file with several dozen stars, and another with orbits of the planets and several asteroids and comets.

1. Create a directory called **AA** on your hard disk and then unzip **aa-54.zip** into it.

2. Exit to DOS, change to the **AA** directory, and then type **AA** to run the program. See **read.me** for additional instructions.

AA200 v. 5.4 (. . PC/AA200)
Contact: Steve Moshier, Email: moshier@na-net.ornl.gov
Type: Freeware
Operating System: DOS

This advanced version of aa-54 computes and displays detailed information about planets, stars, and other bodies using ephemeris data files computed by JPL (not included with the program but available free from http://www.navigator.jpl.nasa.gov).

1. Create a directory on your hard disk called **AA200** and then unzip **aa200-54.zip** into it.

2. Exit to DOS, change to the **AA200** directory; and type **AA200** to run the program. See **readme.aa** for usage instructions.

AAA Eyepiece Analyzer v. 1.02 (. . PC/AAAEPCE)
Contact: Dennis Ward, 347 Jamestown Manor Drive, Gardendale, AL 35071-2646
Type: Demo
Operating System: Windows 3.1 or higher

Enter basic information about the aperture and focal length of your telescope and select a class of eyepiece, and this program displays information on available eyepieces in that category, including

manufacturer's brand and series name, focal length, apparent and actual field of view, power, and size of the exit pupil.

1. Copy all files to a temporary directory on your hard disk and then run **setup.exe** to install the software.

Astrometrica v. 3.1 (. . PC/ASTROMET)
Contact: Herbert Raab, Schrammlstr. 8, A-4050 Traun, Austria; Email: Herbert.Raab@jk.uni-linz.ac.at
Type: Shareware
Operating System: DOS

Conduct astrometric reductions of CCD images to determine the exact positions of objects that appear in the images. The program can read images in FITS format and reduce noise in the images. Select reference stars with known positions to determine the identity and exact location of other objects in the image. Blink, or rapidy flash between two images, to look for objects that may have moved between the two. Requires access to the Hubble Space Telescope Guide Star Catalog (GSC), available separately on CD-ROM, although it can also read alternative star catalogs in particular formats.

1. Create a directory called **ASTROMET** on your hard disk and unzip **astrmt31.zip** into it.

2. Exit to DOS, change to the **ASTROMET** subdirectory, and type **ASTROMET** to run the program. View **readme.txt** for further instructions.

AstroSoft Computerized Ephemeris (. . PC/ASTROSFT)
Contact: Jerry Gardner and Martin Morrison, Astro Soft, Inc., P.O. Box 4451, Hayward, CA 94540
Type: Shareware
Operating System: DOS

This suite of programs provides basic information for observers. ACECALC calculates information for a given location and time, including the rise and set times of the Sun, Moon, and planets and the positions of the Galilean moons of Jupiter, eclipses, equinoxes, and the precession of astronomical coordinates. ACESOLAR is a mini-encyclopedia of information on the Solar System, including basic data on each planet as well as observation tips. ACECAT searches a database of Messier objects, bright stars, and double stars to provide positions, brightness, and commentary on these objects.

1. Create a directory on your hard disk called **ASTROSOFT** and copy the contents of the ASTROSOFT directory into it.

2. Exit to DOS, change to the ASTROSOFT subdirectory, and run the files **ace1.exe** and **ace2.exe** to unpack the software; then enter INSTALL to configure the software for a given location and time.

3. Enter **ACECALC**, **ACESOLAR**, or **ACECAT** to run each program. Documentation is included in the package.

Astrotime v. 2.9 (. . PC/ASTRTIME)
Contact: Bob Nelson, 1393 Garvin St., Prince Geroge, B.C. V2M 3Z1, Canada; Email: nelson@cnc.bc.ca
Type: Shareware
Operating System: Windows 3.1 or higher

Given the time, location, and celestial position of an object, this program provides information on the object's location in the night sky. Can calculate Julian date, sidereal time, air mass and hour angle, and the constellation in which the given coordinates lie. Will also find the azimuthal and zenith angles of an object at a given time, compute the object's galactic longitude and latitude, and convert the coordinates of the object from one epoch to another.

1. Create a directory on your hard disk called **ASTROTIME** and copy all files into it.

2. Run **Astrotime.exe** to start the program.

BRUNGSC 1.00 (. . PC/BRUNGSC)
Contact: Nikolay Sokolov; Email: sokolov@iesl.forth.gr
Type: Freeware
Operating System: DOS

Search for star information stored on Hubble Guide Star Catalog (GSC) CD-ROMs. Specify the center point and size of a chart, as well as the limiting magnitude of stars to be displayed, and the program will display the desired chart on the screen. Requires the GSC CD-ROMs.

1. Create a directory on your hard disk called **BRUNGSC** and copy all files into it.

2. Exit to DOS, change to the BRUNGSC subdirectory, and enter **BRUNGSC** to run the program. See **brungsc.txt** for full directions.

CELL (. . PC/CELL)
Contact: David Chandler, P.O. Box 309, La Verne, CA 91750; Phone: (909) 988-5678
Type: Freeware
Operating System: DOS

A tool for designing telescopes and finding ways to support their mirrors. Enter the diameter, thickness, and focal length of the telescope mirror, and CELL will compute the location and geometry of the support points. Choose among 9, 18, or 27 support points for the telescope.

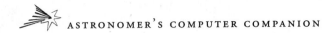

1. Create a directory called **CELL** and copy cell.exe into it.

2. Exit to DOS, change to the **CELL** subdirectory, and type **CELL**. Documentation is included within the program.

CLEA Exercise—Photoelectric Photometry v. 0.74 (.. PC/CLEAPHOT)

Contact: Project CLEA; Phone: (717) 337-6028; Email: clea@ gettysburg.edu
Type: Freeware
Operating System: Windows 3.1 or higher

This educational program simulates observations of stars to measure their brightnesses with a photoelectric detector. Maneuver the telescope to a set of stars in the Pleiades star cluster and use the instrument to measure their magnitudes. Closely simulates an actual telescope, requiring that you center each star in the aperture of the instrument, set the tracking rate correctly, collect data on the sky, and even check that the telescope dome is open! Record the data for analysis outside of the program.

1. Create a directory called **CLEA** on your hard disk and copy all files into it.

2. Change to the **CLEA** subdirectory and run the file **Clea_pho.exe**.

Deep Sky v. 1.4 (.. PC/DEEPSKY)

Contact: Rick Burke, 4234 Cheverage Place, Victoria, B.C. V8N 4Z5, Canada; Phone: (250) 477-4467; Email: burke@octonet.com
Type: Shareware
Operating System: DOS

Helps amateur astronomers identify stars visible in their telescope. Includes a database of 279,000 stars and 16,000 deep-sky objects that can be displayed on the screen. Rotate, invert, and configure the display to match the view from your telescope or search for particular objects.

1. Create a directory called **DEEPSKY** on your hard disk and unzip **deep14-1.zip**, **deep14-2.zip**, and **deep14-3.zip** into it.

2. Exit to DOS, change to the **DEEPSKY** subdirectory, and type **DEEPSKY**. Full instructions and additional installation information is included in the zip files.

Deep*Sky Finder v. 2.0a (. . PC/FINDER)
Contact: Greg Howard, 17 Miron Drive, Poughkeepsie, NY 12603
Type: Freeware ($10 suggested donation)
Operating System: DOS

Search a database of deep-sky objects based on their coordinates, magnitude, associated constellation, catalog name and number, or type (such as spiral galaxy or planetary nebula). Display the results on the screen or print them.

1. Create a directory called **FINDER** on your hard disk and copy **dsWnd20.exe** into it.

2. Exit to DOS, change to the FINDER subdirectory, and run the file dsWnd20.exe to unpack the software. Then type FINDER at the DOS prompt to run the program.

Ephem v. 4.27 (. . PC/EPHEM)
Contact: Elwood Charles Downey; Email: downey@dimed.com
Type: Freeware
Operating System: DOS

Displays the ephemerides, including the right ascension and declination coordinates, distance from the Sun and Earth, and rise and set times for a particular location on Earth, for all the planets and user-defined objects at any given time.

1. Create a directory called **EPHEM** on your hard disk and copy **ephem423.exe** into it.

2. Exit to DOS, change to the **EPHEM** subdirectory, and enter **EPHEM** to run the program. See **man.txt** for more information.

EYEPIECE LE v. 2.70 (. . PC/EYEPIECE))
Contact: Regina L. Roper
Type: Freeware
Operating System: DOS

Use this all-in-one telescope simulator to determine the proper eyepieces and filters to use. Specify the naked-eye limiting magnitude of the night sky and the type of telescope, eyepiece, or filter to calculate key system parameters, including the focal ratio, limiting magnitude of the telescope, field of view, and related information. Choose from an extensive database of existing telescopes and eyepieces or define a custom system.

1. Create a directory called **EYEPIECE** on your hard disk and unzip **eypcle27.zip** into it.

2. Exit to DOS, change to the **EYEPIECE** subdirectory, and type **EYEPCLE** to run the program. Extensive documentation is included with the program.

HBASE (. . PC/HBASE)

Contact: Gary Williams, Rt. 1, Box 138C, Morgan, UT 84050
Type: Freeware
Operating System: DOS

Select a subset from over 2,500 deep-sky objects by specifying a right ascension or declination position, constellation, brightness, object type, or catalog number; then sort or refine your search. Display the results of your search on the screen or print them. The coordinates in the catalog can also be precessed to a particular epoch.

1. Create a directory called **HBASE** on your hard disk and copy **hbase.exe** into it.

2. Exit to DOS, change to the **HBASE** subdirectory, enter **HBASE.EXE** to extract the program, then enter **HBASE** to run it. Documentation is included in the program.

J2000 (. . PC/J2000)

Contact: David Eagle
Type: Freeware
Operating System: DOS

Converts astronomical coordinates from the older B1950 coordinate system to the modern J2000 system. Enter the B1950 coordinates and the proper motion, parallax, and radial velocity of an object to calculate and display its J2000 coordinates and revised proper motion.

1. Create a directory on your hard disk called **J2000** and copy **epoch.exe** into it.

2. Exit to DOS, change to the **J2000** subdirectory, enter **epoch.exe** to extract the program, then enter **J2000** to run it. Documentation is included in the program

KOMSOFT (. . PC/KOMSOFT))

Contact: Stefan Beck, Greutweg 8, 70771 Leinfelden-Echterdingen, Germany
Type: Shareware
Operating System: DOS

Enter comet observations into this program to calculate ephemerides and more. Use KOMSOFT to enter and verify comet observations, providing information about the location, magnitude, and size of the comet as well as the date, time, location, and instrument used for the observation. KOMSOFT will also convert the data into other formats for use by programs such as Guide and Dance of the Planets.

1. Create a directory called **KOMSOFT** on your hard disk and unzip **koms231.zip** into it.

2. Exit to DOS, change to the **KOMSOFT** subdirectory, and type **KOMSOFT** to run the program.

Lunareclipse v. 2.0 (. . PC/LUNECL)

Contact: Christian Nuesch; Email: nuuesch@active.ch
Type: Shareware
Operating System: DOS

Simulates any lunar eclipse visible between the years 1800 and 2300, from any viewing point on Earth. Choose the year, select from among the lunar eclipses visible that year, select a time step and a city from which to observe the eclipse, and run an animation to see the Moon pass through the Earth's shadow. Includes a diary for you to add, edit, and search through reports of eclipse and other observations.

1. Create a directory called **LUNECL** on your hard disk and unzip **eclipse.zip** into it.

2. Exit to DOS, change to the **LUNECL** subdirectory, and type **LUNECL** to run the program. Documentation is included.

ECLIPSE (. . PC/ECLIPSE)
Contact: Luigi Bianchi, 414 Atkinson College, York University, 4700 Keele Street, North York, Ontario M3J 1P3, Canada
Type: Freeware
Operating System: DOS

Provide a starting date and type of eclipse (lunar or solar), and ECLIPSE will search for past or future eclipses. Information displayed includes the date, time, and type of eclipse.

1. Create a directory called **ECLIPSE** on your hard disk and copy **eclipses.exe** into it.

2. Exit to DOS, change to the **ECLIPSE** subdirectory, and enter **ECLIPSES** to unpack the software; enter **ECLIPSE** to run the program. See **eclipse.doc** for further information.

Messier Logging System (MLS) v. 1.0 (. . PC/MESSIER)
Contact: Luis Argüelles; Email: whuyss@tripod.net; Web: http://members.tripod.com/~whuyss/mls.htm
Type: Freeware
Operating System: Windows 95 or Windows NT

This free, customizable database of the Messier objects, provides the Messier ID number, associated constellation, celestial coordinates (right ascension and declination), a short description, magnitude, and various additional information for each object in the database. Includes an image of each Messier object for ease of identification.

1. Run **setup.exe** from the /MESSIER/Disk1 folder on the CD-ROM to install.

Moonrise (.. PC/MOONRISE)

Contact: Bruce Sidell, 3115 Cascade SE, Grand Rapids, MI 49506;
Email: bsidell@iserv.net; Web: http://www.iserv.net/~bsidell/
moonrise.htm
Type: Shareware
Operating System: Windows 95 or higher

Computes the rise and set times of the Sun and the Moon for any
given day and displays an icon of the current Moon phase. Search for
recent and upcoming Moon phases. Note: Requires the Visual Basic
4.0 runtime files (see /PC/vbrun400).

1. Run **setup.exe** to install.

NGCView (.. PC/NGCVIEW)

Contact: Rainman Software, P.O.Box 6074, Gloucester, MA 01930-
6174; Sales: (888) 261-1686; Fax: (888) 995-9191; Email: ngcview@
rainman-soft.com; Web: http://www.rainman-soft.com
Type: Demo
Operating System: Windows 3.1 (demo16.exe) and Windows 95 or
higher (51demo32.exe)

NGCView is an outstanding observation tool that combines many of
the best qualities of planetarium, ephemerides, and text-based database programs to optimize observation time. Its four primary
components, Find It, Sort It, Chart It, and Log It, each offer great
depth and excellent design. NGCView is highly customizable and
displays and prints very high-quality charts.

1. Run either **demo16.exe** (Windows 3.1) or **51demo32.exe**
   (Windows 95 or higher) to install.

OBSERVE (. . PC/OBSERVE)

Contact: Weems S. Hutto, 2704 El Greco Lane, Dallas, TX 75287;
Phone: (214) 306-6981
Type: Freeware
Operating System: DOS

Organize objects for planned observing sessions and retrieve information about deep-sky objects from catalogs. Search deep-sky objects by Messier, NGC, or Herschel catalog number for their locations, sizes, and magnitudes and enter observational notes including date, time, instrument used, and observation conditions for later reference.

1. Create a directory on your hard disk called **OBSERVE** and copy **observer.exe** into it.

2. Exit to DOS, change to the **OBSERVE** subdirectory, enter **OBSERVER** to unpack the program then enter **OBSERVE** to run it.

OPTICS v. 1.0 (. . PC/OPTICS)

Type: Freeware
Operating System: DOS

Enter the aperture, mirror focal length, eyepiece type, and focal length to calculate a number of key parameters for a telescope system, including the effective f-stop, limiting magnitude, telescope power, and more. Given the field diameter and a step size, the program will also compute the astigmatism of the system.

1. Create a directory on your hard disk called **OPTICS** and copy **optics1.exe** into it.

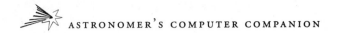

2. Exit to DOS, change to the **OPTICS** subdirectory, run **optics1.exe** to unpack the software, and then enter **OPTICS** to run the program.

REDSCOPE v. 4.04 (. . PC/REDSCOPE)
Contact: Regina L. Roper
Type: Freeware
Operating System: DOS

Calculate telescope and eyepiece performance (including magnification, limiting magnitude, and resolution), given the telescope aperture and focal length, eyepiece specifications, and other details about a telescope system. Select from a set of deep-sky object databases, including the Messier objects, or lists of double stars, to determine how easily the object can be observed with the specified telescope system.

1. Create a directory on your hard disk called **REDSCOPE** and unzip **red404.zip** into it.

2. Exit to DOS prompt, change to the **REDSCOPE** subdirectory, and enter **REDSCOPE** to run the program. See **redscp.doc** for further instruction.

SatFlash v. 5.1 (. . PC/SATFLASH)
Contact: Belgian Working Group Satellites, Bart de Pontieu, Vinkenstraat 23, B-8400 Oostende, Belgium
Type: Freeware
Operating System: DOS

Track and analyze the flash periods of artificial satellites (the times when they appear to suddenly brighten). Process imported observations, add new ones, or study recent ones. Also provides information on observation reports and calculates the physical elements of orbiting satellites.

1. Create a directory on your hard disk called **SATFLASH** and copy **sf51.exe** into it.

2. Exit to DOS, change to the **SATFLASH** subdirectory, and enter **sf51.exe** to unpack the software. Enter **SF** to run the software.

SIDEREAL (. . PC/SIDEREAL)
Contact: David Eagle
Type: Freeware
Operating System: DOS

Converts between several different astronomical date and time systems, including normal calendar and Julian dates (the number of days since January 1, 4713 BC). Enter a specific longitude to convert between the local sidereal and the local civial time.

1. Create a directory on your hard disk called **SIDEREAL** and copy **sdreal.exe** into it.

2. Exit to DOS, change to the **SIDEREAL** subdirectory, and enter **SDREAL** to unpack the software. Enter **SIDEREAL** to run the program.

SIMON 1.0 (. . PC/SIMON)
Contact: Stefanik Observatory, Petrin 205, Praha 1, 118 46, Czech Republic; Email: lenka@asu.cas.cz; Web: http://sorry.vse.cz/~ludek/zakryty/pub.html
Type: Freeware
Operating System: DOS

Simulates stellar occultations by the Moon. A graphic of the Moon and a star are displayed, and the user can adjust the phase, colors, brightness, and other aspects of the Moon. SIMON also allows users to test their timing reflexes: the star disappears behind the Moon, and you must press the ENTER key as soon as it does. The program then displays the reaction time (the difference between the time the star disappeared and the time you pressed the key). This is a way to help train your reflexes to more accurately record the times of an occultation by eye.

1. Create a directory on your hard disk called **SIMON** and unzip **simon10a.zip** into it.

2. Exit to DOS, change to the **SIMON** subdirectory, and enter **SIMON** to run the program. Use the **F10** key, if necessary, to toggle the controls to read in English. See **simon.doc** for further information.

SKY-ATLAS v. 1.4 (. . PC/SKYATLAS)
Contact: TomiSoft, Tomi Heinonen, Meriraumantie 20D, 26200 Rauma, Finland; Email: tomi@cc.tut.fi
Type: Shareware
Operating System: DOS

This program draws star charts of a desired region of the sky using the SAO (Smithsonian Astrophysical Observatory) catalog of over 100,000 stars. SKY-ATLAS can display stars down to magnitude 11 and generate equatorial or north and south polar maps, which you can zoom in on for more detail. It also displays deep-sky objects such as galaxies and nebulae and offers information on them in its database.

1. Create a directory on your hard disk called **SKYATLAS** and copy all files into it.

2. Exit to DOS, change to the **SKYATLAS** subdirectory, and type **ESAO** to run the program.

SkyBase 2000.0 Skychart v. 4.31 (. . PC/SKYBASE)

Contact: Logan Zintsmaster, 450 Spring Hill Dr., Morgan Hill, CA 95037
Type: Shareware
Operating System: DOS

Displays several types of star charts for a given time and location. Includes the Yale Bright Star Catalog, SAO catalog, and Messier and some NGC objects, all down to magnitude 9. You can specify the data files from which to obtain the star and deep-sky object positions, the center and size of the map to be plotted, and whether it should be inverted and/or mirrored. You can print maps or zoom in on them for more detail.

1. Create a directory on your hard disk called **SKYBASE** and copy **skyb431.exe** into it.

2. Exit to DOS, change to the **SKYBASE** subdirectory, and enter **SKYB431** to unpack the software. Enter **SKYBASE** to run the program.

STARPLOT v. 2.0 (. . PC/STARPLOT)

Contact: Bernie Czenkusz, 3 Briggs Grove, Wibsey, Bradford, West Yorkshire, England
Type: Shareware
Operating System: DOS

Create, edit, and print custom star charts for specific regions of the sky using STARPLOT's catalog of stars or the Hubble Space Telescope Guide Star Catalog. Maneuver around the sky with your mouse to read the right ascension and declination for any position on the screen. Includes Makeplan, which creates planet, comet, and asteroid charts for use in the main program.

1. Create a directory called **STARPLOT** and unzip **starpl20.zip** into it.

2. Exit to DOS, change to the **STARPLOT** subdirectory, and enter **STARPLOT** to run the program or **README** to read the documentation.

Teleodm (. . PC/TELEODM)

Contact: David Lutz; Email: lutz@scp.rochester.edu
Type: Freeware
Operating System: DOS

Enter information about your telescope and viewing conditions, and Teleodm will tell you whether a particular deep-sky object will be visible with your telescope. The program prompts for the telescope aperture and sky brightness and lists the magnitude of the faintest star observable.

1. Create a directory on your hard disk called **TELEODM** and unzip **odmv11b.zip** into it.

2. Exit to DOS, change to the **TELEODM** subdirectory, and enter **TELEODM** to run the program. Background information on the application is included in the zip file.

Telescope Calc v. 1.1e (. . PC/TELESCOP)

Contact: Stephan Schilling; Email: sschill@imn.htwk-leipzig.de
Type: Shareware
Operating System: Windows 3.1 or higher

Given the telescope diameter and the focal length of the telescope and eyepiece, along with the age of your eye pupil, Telescope Calc will calculate the normal and maximum useful magnification, the magnification with the given eyepiece, the size of the eyepiece pupil, the best possible angular separation, and the minimum brightness than can be observed with this telescope and eyepiece system.

1. Create a directory on your hard disk called **TELESCOP** and copy all files into it.

2. Run the program **tel_eng.exe**.

ULTRA TRACE (. . PC/ULTRA)

Contact: NOVA Optical Systems, PO Box 80062, Cornish, UT 84308; Email: nova@cache.net; Web: http://www.astronomy-mall.com
Type: Shareware
Operating System: Windows 3.1 (ULTRA), Windows 95 or higher (ULTRA32)

This ray-tracing program can help you design the optics for a telescope system. Enter the basic characteristics of a telescope, such as diameter and focal length, and the program computes a number of more specific parameters, including limiting magnitude, maximum magnification, theoretical resolution, size of any central obstruction, and more. Then run a ray-tracing calculation to show the size of the Airy disk created by light coming from off-axis at a specified wavelength. (You'll need to register the software to design anything other than a Newtonian telescope.)

1. Run **setup** in the appropriate directory on the CD-ROM to install (use the **ULTRA** directory for Windows 3.1 systems, **ULTRA32** for Windows 95 or higher and Windows NT).

Photography Tools

Astrophotography Exposure Calculator (. . PC/ASTROEXP)
Contact: Michael Covington; Web: http://www.mindspring.com/~covington/astro
Type: Freeware
Operating System: Windows 95 or higher

Calculates exposures for astronomical photographs.

   1. Run **ASTROEXP.exe.**

Planetarium Programs

ASTRO version 2.02 (. . PC/ASTRO202)
Contact: Athabasca University, Alberta, Canada; Email: reeves@rocky1.usask.ca
Type: Freeware
Operating System: DOS

Given a date and time, as well as the latitude and longitude of the observer, this simple planetarium program will generate a map of the sky, including the positions of the Sun, Moon, planets, and several hundred stars from the Yale Bright Star Catalog. The star displays include information on the magnitude and spectral class of each star and can be saved as GIF files or exported to PostScript and other formats.

   1. Create a directory on your hard disk called **ASTRO202** and copy all files into it.

   2. Exit to DOS, change to the **ASTRO202** subdirectory, and enter **ASTRO** to run the program. See **astro.doc** for usage information.

Coeli—Electric Planisphere v. 3.85 (. . PC/COELI)

Contact: Roger Hughes, Swimming Elk Software; Email: Swimming.
Elk@sci.fi; Web: http://www.sci.fi/~elk
Type: Shareware
Operating System: DOS/Windows 3.1 (COELI/COELI385), Windows 95 or higher (COELI/STELLA12)

This is an advanced planetarium program for viewing the night sky from any location or time. The software includes thousands of stars down to the naked-eye limit, with information about each that includes coordinates and rise and set times. An included world map makes it easy to identify a particular observation location.

DOS/Windows 3.1:

1. Create a directory on your hard disk called **COELI** and copy all files from /COELI/COELI385 into it.

2. Exit to DOS, change to the COELI subdirectory, and run **coelinst.exe** to unpack the files. Enter **COELI** to run the program. (Consult the file **install.txt** for complete instructions on installing the software.)

Windows 95 or higher:

1. Run **setup.exe** from the /COELI/STELLA12 subdirectory on the CD-ROM.

Earth Centered Universe v. 3.0A (. . PC/ECUSHARE)

Contact: Nova Astronomics, David J. Lane, PO Box 31013, Halifax, Nova Scotia, B3K 5T9 Canada; Phone: (902) 499-6196; Fax: (902) 826-7957; Email: info@nova-astro.com; Web: http://www.nova-astro.com
Type: Shareware
Operating System: Windows 3.1 or higher

A moderately priced planetarium and telescope control program with a database of over 15,000,000 stars, the planets, Sun, Moon, comets, over 30,000 asteroids, and more than 10,000 deep-sky objects. Prints high-quality star charts using any Windows-compatible printer and controls modern computerized telescopes such as the Meade LX200 series.

1. Run **ecushare.exe** to install.

HUBBLE Space Telescope v. 1.91 (. . PC/HUBBLE)
Contact: Sci-Vision, P.O. Box 941532, Suite 1159, Dallas, TX 75094
Type: Demo
Operating System: Windows 3.1 or higher

This simple planetarium program can display the night sky, as well as the Solar System and Jupiter's Galilean satellites, for a given time and location. Click on an object in the star chart to display additional information about it. A special feature lets you compare the relative sizes of different planets, moons, and stars by plotting them side by side on the screen.

1. Create a directory on your hard disk called **HUBBLE** and copy all files into it.

2. Run the program **Hubble.exe** in the **HUBBLE** directory. A help file included in the zip file has extensive documentation on the program.

MyStars! v. 2.4.1 (. . PC/MYSTARS)
Contact: David Patte, Relative Data Products, 365 Sherwood Drive, Ottawa, Ontario, Canada K1Y 3X3; Fax: (613) 728-4240; Email: dpatte@relativedata.com; Web: http://www.relativedata.com/
Type: Shareware
Operating System: Windows 3.1 (16-BIT), Windows 95 or higher (32-BIT)

This basic planetarium program displays stars, planets, comets, and deep-sky objects, and information about them. Shows actual planet images to scale and an orrery of planets in their orbits around the Sun. The display can be zoomed in and out and even animated over time.

1. Run **install.exe** in the appropriate MYSTARS directory (see above). A help file is included with detailed information about how to use the program.

Sky View v. 1.15d (. . PC/SKYVIEW)

Contact: M. Turini, Via Piantalbis 109, 56023 Cascina (PI), Italy; Phone: 39-50-771220; Fax: 39-50-744267; Email: sinclair@cli.di.unipi.it
Type: Shareware
Operating System: DOS

Displays the night sky for any night between 32,000 BC and 32,000 AD, including stars, planets, and deep-sky objects. Click on an object with the right mouse button to bring up a window of information about the object, including name, brightness, coordinates, rise and set times, and more. The left mouse button brings up a list of options.

1. Create a directory on your hard disk called **SKYVIEW** and unzip **sv115.zip** into it.

2. Exit to DOS, change to the **SKYVIEW** subdirectory, and enter **SV** to run the program.

Sky3D v. 1.1 (. . PC/SKY3D)

Contact: Corvus Software, Eduardo Nunes, R. Sabino de Sousa,17 2Dt, 1900 Lisboa, Portugal; Email: l40690@alfa.ist.ult.pt; Web: http://www.getsoftware.com/corvus
Type: Shareware
Operating System: DOS

Zoom in and out of the sky and get information on stars and galaxies using three viewing modes that enable 3D viewing of objects from any vantage point in the galaxy. Includes 9,100 stars, with information available about each when you select them with the mouse. Export 3D maps into formats suitable for ray-tracing programs like POV-Ray or in VRML for the Web.

1. Create a directory on your hard disk called **SKY3D** and copy all files into it.

2. Exit to DOS, change to the **SKY3D** directory, and enter **SKY3D** to run the program. See **readme.txt** for additional documentation.

SkyGlobe v. 3.6 (. . PC/SKYGLOBE)

Contact: Mark Haney, KlassM Software, Inc., PO Box 1067, Ann Arbor, MI 48106; Email: via CompuServe, 75020,1431; Phone: (800) 968-4994
Type: Shareware
Operating System: DOS

SkyGlobe is an educational planetarium program. Use the mouse to rotate around a view of the night sky; center the crosshairs over an object to display its name; and click the left mouse button to center the display on that object. Select a location from a list of over 200 cities. The program can display as many as 29,000 stars down to magnitude 7.6, and you can jump forward a minute, hour, day, month, year, or millennium at a time to see how the sky changes over time.

1. Create a directory on your hard disk called **SKYGLOBE** and copy all files into it.

2. Exit to DOS, change to the **SKYGLOBE** subdirectory, and enter **SKYGLOBE** to run the program.

SkyMap Pro v. 5.0 (. . PC/SKYMAP)

Contact: SkyMap, 9 Severn Road, Culcheth, Cheshire WA3 5ED, UK; Phone: 44-1925-764131; Technical support: World Wide Software Publishing, PO Box 202036, Bloomington, MN 55420; Phone: (612) 881-7045; Fax: (612) 881-7024; Email: support@wwsoftware.com or sales@skymap.com; Web: http://www.skymap.com
Type: Demo
Operating System: Windows 95 or higher

This demonstration version of SkyMap Pro 5 is 6.6MB in size, and comprises the fully functional program and manual, together with a small database of 30,000 stars (down to magnitude 7.5) and 10,000 deep sky objects. It runs entirely from your hard disk, occupying 13MB of hard disk space when installed. The full version of SkyMap Pro can display more than 15 million stars and 100,000 deep-sky objects.

1. Run **smp5.exe** to install the program.

SKYPLOT v. 2.0 (. . PC/SKYPLOT)

Contact: Gerald M. Santoro, 148 W. Hamilton Ave., State College, PA 16801
Type: Shareware
Operating System: DOS

This basic planetarium simulator can display a basic map of the northern, southern, eastern, or western sky at a given position (in latitude and longitude) for any time between the years 1975 and 2500. The program can use stars down to magnitude 5 from the Yale Bright Star Catalog and displays the positions of Solar System objects. It can also plot objects from a user-supplied file of objects and positions.

1. Create a directory on your hard disk called **SKYPLOT** and copy all files into it.

2. Exit to DOS, change to the **SKYPLOT** subdirectory, and enter **SKYPLOT** to run the program. See **skyplot.doc** for further documentation.

Starry Night Basic v. 2.1.2 (. . PC/STRYNITE)

Contact: Sienna Software, Inc., 411 Richmond Street East, Suite 303, Toronto, Ontario, Canada, M5A 3S5; Sales: (800) 252-5417; Phone: (416) 410-0259; Fax: (416) 410-0359; Email: contact@siennasoft.com; Web: http://www.siennasoft.com
Type: Shareware
Operating System: Windows 95 or higher

View the night sky from any place on Earth or in the Solar System at any time. A "Swiss Army Knife" Astronomy application, Starry Night is very feature rich and is being used around the world by parents, educators, amateur astronomers, and those just beginning to discover the wonders of astronomy. The deluxe CD-ROM version adds many features, including a greatly expanded database of stars and of comets whose tails grow in size as they approach the Sun; the Macintosh version offers a plug-in that can control computerized telescopes.

1. Run **snbas212.exe** to install.

Note: Starry Night Basic requires QuickTime 3.x for full functionality; a copy of QuickTime 4.0 is included on this CD-ROM in the PC/QTIME directory.

StarScape v. 1.54 (. . PC/STARSCPE)

Contact: Skyline Software, J M Technical Services, 11a Cedar Lane, Cockermouth, Cumbria CA13 9HN, UK; Email: jmtech@netlink.co.uk; Web: http://www.netlink.co.uk/users/jmtech
Type: Freeware
Operating System: Windows 3.1 or higher

Displays the positions of stars and planets in the night sky for a particular location and time, including constellation names and lines and deep-sky objects. Click on an object for basic information about it or search for a particular object.

1. Create a directory on your hard disk called **STARSCP** and copy all files into it.

2. Run **install.exe** from the **STARSCP** directory to install the program.

StarView v. 2.0.08 (. . PC/STARVIEW)

Contact: Brian McMillin, Digi-Tech Consultants, P. O. Box 12144, Dallas, TX 75225
Type: Shareware
Operating System: DOS

Displays the locations of stars, planets, and other objects in the night sky for a given location and time. Simulates the view from a telescope, including image inversion and tracking. By clicking and dragging the mouse, you can move around the sky, adjust the size of the field, and change the number of stars visible (the program includes over 9,100 stars down to magnitude 9). Center the view by entering the name of an object, its position, or its catalog number. Add your own objects to the database using a separate text file described in the program's help section.

1. Create a directory on your hard disk called **STARVIEW** and copy all files into it.

2. Exit to DOS, change to the **STARVIEW** subdirectory, and enter **STARVIEW** to run the program.

The Big Blue Skies Planetarium (. . PC/BLUESKY)
Contact: Ichiroh Sasaki
Type: Freeware
Operating System: DOS

Display the sky at any given time and location and zoom in on specific constellations for additional information. Choose from a database of 1,500 or over 3,000 stars or display a single constellation and get information on the positions and magnitudes of the stars within it.

1. Create a directory on your hard disk called **BLUESKY** and copy **blsky12a.exe** and **blsky12b.exe** into it.

2. Exit to DOS, change to the **BLUESKY** subdirectory, and run **blsky12a.exe** and **blsky12b.exe** to unpack them.

3. Enter **BLUESKY** to run the program.

Satellite Programs

FLYBY (. . PC/FLYBY)
Contact: Science Software; Email: 74561.606@compuserve.com
Type: Freeware
Operating System: DOS

Calculates the parameter of spacecraft flybys of planets for gravity-assists using a data file that specifies the starting and ending planet and any planets in between to be used for flybys, as well as the launch date and the flyby altitude. FLYBY steps through a set of possible solutions and, when it finds a viable one, it displays detailed parameters of the solution, including the flyby and arrival dates and turn angles.

1. Create a directory on your hard disk called **FLYBY** and copy all files into it.

2. Exit to DOS, change to the **FLYBY** subdirectory, and enter **FLYBY** to run the program.

GENTRA v. 1.0 (. . PC/GENTRA)

Contact: David Eagle
Type: Freeware
Operating System: DOS

Use this program for interactive orbit analysis, converting among several different orbital and time formats and solving the Kepler equation for elliptical orbits. GENTRA will also calculate information about the orbit, such as the period, perigee, and apogee. The program can read data from separate data files and write results to data files.

1. Create a directory on your hard disk called **GENTRA** and copy all files into it.

2. Exit to DOS, change to the **GENTRA** subdirectory, and enter **GENTRA** to run the program.

GRMOTION (. . PC/GRMOTION)

Contact: None
Type: Freeware
Operating System: DOS

Computes the relative distance between two orbiting satellites. Enter a set of initial conditions about an orbit, including the altitude of the target satellite and the initial position and velocity of the secondary satellite, or calculate and display rendezvous or synchronous orbits to calculate the relative motion between the two satellites.

1. Create a directory on your hard disk called **GRMOTION** then copy **grmotion.exe** into it.

2. Exit to DOS, change to the **GRMOTION** subdirectory, and enter **GRMOTION** to run the program.

IPTO (. . PC/IPTO)
Contact: Science Software; Email: 74561.606@compuserve.com
Type: Freeware
Operating System: DOS

Computes an optimized trajectory for a spacecraft traveling between planets. This "patched conic" trajectory can be computed to minimize the launch and/or arrival energy of a spacecraft following it. The parameters for the orbit, including planet names and launch and arrival dates, are read from a separate text file. Displays the orbit results, including best launch and arrival dates, travel time, energy, and orbital elements.

1. Create a directory on your hard disk called **IPTO** and copy all files into it.

2. Exit to DOS, change to the **IPTO** subdirectory, and enter **IPTO** to run the program or **README** for more information.

KEPLER (. . PC/KEPLER)
Contact: David Eagle
Type: Freeware
Operating System: DOS

Use KEPLER to solve the challenging Kepler Equation for the motion of a satellite in orbit around another body. Enter the mean anomaly of an object's orbit and the orbit's eccentricity to compute the eccentric anomaly for that orbit.

1. Create a directory on your hard disk called **KEPLER** and **copy**
    kepler1.com into it.

2. Exit to DOS, change to the **KEPLER** subdirectory, run
    kepler1.com to extract the program, and then enter **KEPLER**
    to run the program.

LogSat Professional v. 5.2 (. . PC/LOGSAT)

Contact: LogSat Software Corporation, 425 S.Chickasaw Tr., Suite
103, Orlando, FL 32825; Phone: (800) 350-3871; International: +1 407-
275- 0780; Fax: (407) 275-2416; Email: sales@logsat.com; Web: http://
www.logsat.com
Type: Demo
Operating System: Windows 3.1 or higher

Track thousands of satellites (2,000 to a window) with LogSat—even
open multiple windows and assign each its own time and date. A digitized voice will even notify you of events. Five different map
projections are available—Mercator, equidistant, sinusoidal, Hammer, and orthographic—and LogSat will display 3D radiation
diagrams for antennas and graphs on ground-wave propagation.

1. Run **logsat52.exe** from the CD-ROM and follow the onscreen
    instructions.

Orbit Simulation v. 4.0 (. . PC/ORBIT)

Contact: USAF Academy Department of Astronautics, Attn: Direc-
tor of Education, Technology, USAFA/DFAS, USAF Academy, CO
80840; Phone: (719) 472-4110
Type: Freeware
Operating System: DOS

Enter the standard orbital elements that define the size, shape, and orientation of an orbit to simulate the orbits of satellites around the Earth. Select a viewing orientation from which to watch the orbit to display the orbit from that vantage point as well as a ground track (the path the satellite appears to follow over the Earth) and updated orbital parameters.

1. Create a directory on your hard disk called **ORBIT** and copy **orbit.exe** into it.

2. Exit to DOS, change to the **ORBIT** subdirectory, and enter **ORBIT** to run the program.

PC-TRACK v. 3.1 (. . PC/PCTRACK)

Contact: Thomas C. Johnson, 9920 S. Palmer Rd., New Carlisle, OH 45344; Phone: (513) 878-6030
Type: Shareware
Operating System: DOS

Easily track the location of Earth-orbiting spacecraft based on satellite elements provided by NASA and other organizations. Track up to 200 satellites simultaneously and plot their locations on a world map in 3D for any given date and time. PC-TRACK will even point out which objects are illuminated and thus would be visible in the night sky. The program can read data files in standard NASA two-line orbital element format to keep it up to date.

1. Create a directory on your hard disk called **PCTRACK** and copy all files into it.

2. Exit to DOS, change to the **PCTRACK** subdirectory, and enter **PCT3** to run the program.

Space Flight Simulator v. 1.01 (. . PC/SFS)

Contact: Bywater Software, P. O. Box 4023, Duke Station, Durham, NC 27707; Email: tcamp@teer3.acpub.duke.edu
Type: Freeware
Operating System: DOS

A graphical simulation of space flight, with real-time, low-resolution animations of spacecraft orbiting the Earth and other bodies. View the ground track produced by an orbiting spacecraft as well as the view of the Earth from the spacecraft itself or a view from an external vantage point. The program includes a number of test cases, including the orbit of the Hubble Space Telescope around the Earth as well as orbits around the Moon and Mars. The parameters of the orbit, including its height, eccentricity, and inclination, can all be modified within the program.

1. Create a directory on your hard disk called **SFS** and unzip **sfs101.zip** into it.

2. Exit to DOS, change to the **SFS** subdirectory, and enter **SFS** to run the program. Additional documentation is included in the **readme.txt** file.

STS26 (. . PC/STS26)

Contact: Robert Lloyd
Type: Shareware
Operating System: DOS

Given information on the Space Shuttle's orbit, STS26 plots its location above the Earth at any given time using a world map. The program asks for the launch direction, inclination, and altitude or period of the orbit, and a starting time (in days, hours, and minutes since launch). The plot can be run forward in time to see the orbital track of the Space Shuttle.

1. Create a directory on your hard disk called **STS26** and copy **sts.com** into it.

2. Exit to DOS, change to the **STS26** subdirectory, and run **sts. com** to unpack the software. Enter **STS26** to run the program.

STSORBIT PLUS v. 9848 (. . PC/STSPLUS)

Contact: David H. Ransom, Jr., 240 Bristlecone Pines Road, Sedona, Arizona 86336; Email: rans7500@spacelink.nasa.gov; Web: http:// www.dransom.com
Type: Shareware
Operating System: DOS

Widely used throughout NASA and the USAF, STSORBIT PLUS is intended for use during Space Shuttle missions and for general satellite tracking using NASA/NORAD two-line orbital elements. Track one primary and up to 32 additional static or real-time satellites simultaneously, in real time. Both orthographic and rectangular VGA color map projections are available, displaying the Earth as a globe or the more traditional flat map. A Night Vision mode shows the maps in red to protect your night vision. Predicts and displays tabular line-of-sight satellite passes.

1. Create a directory on your hard disk called **STSPPLUS** and copy the contents of the CD-ROM's **STSPLUS** directory into it.

2. Exit to DOS, change to the **STSPLUS** subdirectory, and unzip all files.

3. Enter **STSPLUS** to be guided through a configuration menu.

TEKOSIM v. 2.16a (. . PC/TEKOSIM)

Contact: Leo Wikholm, Keinutie 7 F 139, SF-00940 Helsinki, Finland
Type: Freeware
Operating System: DOS

Simulates the motions of satellites against the night sky to give you practice with satellite crossing observations. As the satellite moves across the screen, press **ENTER** to make it stop and disappear. Then move the crosshairs to where you think it was just before it disappeared, to test your accuracy and practice for actual satellite observations.

1. Create a directory on your hard disk called **TEKOSIM** and copy **ts216a.exe** into it.

2. Exit to DOS, change to the **TEKOSIM** subdirectory, and run **ts216a.exe** to unpack the software. Enter **TS** to run the program.

TRAKSAT v. 4.08 (. . PC/TRAKSAT)

Contact: Paul E. Traufler, 111 Emerald Dr., Harvest AL 35749; Phone: (256) 837-0084; Fax: (256) 895-0754; Email: wintrak@traveller.com; http://www.hsv.tis.net/~wintrak
Type: Shareware
Operating System: DOS

TRAKSAT uses the two-line elements of satellite information provided by NORAD to determine where and when a satellite is visible from a specified location. The experienced satellite tracker will like all the power and flexibility TRAKSAT provides.

1. Create a directory on your hard disk called **TRAKSAT** and copy all files into it.

2. Exit to DOS, change to the **TRAKSAT** subdirectory, and enter **TRAKSAT** to run the program. See **traksat.doc** for additional information.

Screen Savers

CosmoSaver-Solar System v. 1.31 (. . PC/COSMO)
Contact: MicroRealities, P.O. Box 2756, Sudbury, Ontario, P3A 5J3, Canada; Email: support@microrealities.com; Phone: (705) 524-9331
Type: Shareware
Operating System: Windows 3.1 or higher

This set of astronomical screen savers includes 3D images of the Sun, planets, and moons. The screen savers allow you to see moons orbiting each planet or to touch down on the surface of the planet to see the night sky (including blue daytime skies on Earth). The animations for each planet can be individually configured.

1. Run **Setup.exe** to install.

Craters Screen Saver (. . PC/CRATERS)
Contact: John Walker, Fourmilab; Web: http://www.fourmilab.ch
Type: Freeware
Operating System: Windows 95 or higher

A screen saver that simulates the cratering of initially flat terrain, obeying the same power law that relates crater size to number observed on airless Solar System bodies.

1. Copy the **craters.scr** file to your main Windows directory (usually C:\WINDOWS). Use the Windows **Control Panel** to select and configure the screen saver. (See your Windows documentation for help.)

Earth Screen Saver (. . PC/EARTH)

Contact: John Walker, Fourmilab; Web: http://www.fourmilab.ch
Type: Freeware
Operating System: Windows 95 or higher

This screen saver shows the Earth with the correct illumination based on the date and time. You can view Earth from the Sun (day side) or Moon (night side) or at a given altitude above any location.

1. Copy the **earth.scr** file to your main Windows directory (usually C:\WINDOWS). Use the Windows **Control Panel** to select and configure the screen saver. (See your Windows documentation for help.) See **readme.txt** for additional information.

Sky Screen Saver (. . PC/SKY)

Contact: John Walker, Fourmilab; Web: http://www.fourmilab.ch
Type: Freeware
Operating System: Windows 3.1 (SKYSCRSN), Windows 95 or higher (SKYSCR32)

This screen saver shows the sky, including stars from the more than 9,000-star Yale Bright Star Catalog; the Sun, Moon (with the correct phase), and planets; deep-sky objects drawn from a database of more than 500 prominent objects including all Messier objects; constellation names, boundaries, and outlines; and ecliptic and equatorial coordinates. All items can be individually selected to customize the display, and the screen saver can be configured for any time zone and any location on Earth.

1. Copy the appropriate **.scr** file to your main Windows directory (usually C:\WINDOWS). Use the Windows **Control Panel** to select and configure the screen saver. (See your Windows documentation for help.) See **readme.txt** for further information.

Solar System Programs

CLEA Exercise—Mercury Rotation v. 0.70 (. . PC/CLEAM)
Contact: Project CLEA, Phone: (717) 337-6028; Email: clea@ gettysburg.edu
Type: Freeware
Operating System: Windows 3.1 or higher

This educational program uses simulated radar observations of Mercury to determine its rotation rate. Send out radar pulses and measure their return, looking for a frequency shift caused by Mercury's rotation and its curved surface; then use that data to compute the rotation rate of the planet.

1. Create a directory on your hard disk called **CLEAM** and unzip **clea_mer.zip** into it.

2. Change to the **CLEAM** subdirectory and run **clea_mer.exe**.

CLEA Exercise—Moons of Jupiter v. 0.75 (. . PC/CLEAJ)
Contact: Project CLEA; Phone: (717) 337-6028; Email: clea@ gettysburg.edu
Type: Freeware
Operating System: Windows 3.1 or higher

This educational program simulates observations of the Galilean moons of Jupiter by having you study a telescopic image of the moons of Jupiter and measuring the positions of the moons relative to Jupiter over a series of nights. Once you've collected enough data, you compute the orbits of each of the four moons (which must be done outside of the program).

1. Create a directory on your hard disk called **CLEAJ** and copy all files into it.

2. Run **clea_jup.exe**.

EarthSun v. 2.5 (. . PC/EARTHSUN)

Contact: W. Scott Thoman, 41 Lee Road, Dryden, NY 13053; Email: thoman@law.mail.cornell.edu
Type: Shareware
Operating System: Windows 3.1 or higher

Automatically generates an icon showing the sunlit face of the Earth at the current time. The icon shows the current time and can adjust to Daylight Savings Time manually or automatically, depending on the country selected.

1. Create a directory on your hard disk called **EARTHSUN** and copy all files into it.

2. Run **earthsun.exe**. Consult the help file (**earthsun.hlp**) included with the software for full instructions.

Home Planet v. 2.1 (. . PC/HOMEPLAN/21)

Contact: John Walker; Email: kelvin@fourmilab.ch; Web: http://www.fourmilab.ch
Type: Freeware
Operating System: Windows 3.1 or higher

Generates a high-resolution map of the Earth and shows the portions of the Earth that are in sunlight at a given time. View the Earth from the Sun, Moon, night side, or orbiting spacecraft, and plot the stars and planets visible above a certain location with features on a par with many specialized planetarium programs. Home Planet can also display an orrery (the positions of the planets in their orbits around the

Sun); calculate current information on the location of the Sun and the Moon; and track satellites and display them on the map of the Earth, using an extensive built-in database.

1. Run **Setup.exe** to install

Home Planet v. 3.0 (. . PC/HOMEPLAN/30)
Contact: John Walker; Email: kelvin@fourmilab.ch; Web: http://www.fourmilab.ch
Type: Freeware
Operating System: Windows 95 or higher

This comprehensive astronomy, space, and satellite-tracking package also calculates the current location of the Sun and Moon.

1. Run **hplanet.exe** from the CD-ROM or copy the folder to your hard disk.

Jovian Satellites' Simulator v. 4.0 (. . PC/JOVIAN)
Contact: Gary Nugent, 54a Landscape Park, Churchtown, Dublin 14, Ireland; Email: gnugent@cara.ie
Type: Shareware
Operating System: DOS

Calculates and displays the locations of Jupiter's Galilean satellites (Io, Europa, Ganymede, and Callisto) at a given time. The display can be animated to show the motions of the satellites over time, including plotting lines to show the tracks of the satellites' orbits, or altered to show how the view might appear through a telescope, binoculars, or other instruments.

1. Create a directory on your hard disk called **JOVIAN** and copy all files into it.

2. Exit to DOS, change to the **JUPSAT** subdirectory, and enter **JUPSAT** to run the program. See more detailed instructions in **jupsat.txt**.

JupSat95 v. 1.14 (. . PC/JUPSAT)

Contact: Gary Nugent, 54a Landscape Park, Churchtown, Dublin 14, Ireland; Email: gnugent@indigo.ie; Web: http://indigo.ie/~gnugent/ JupSat95
Type: Shareware
Operating System: Windows 95 or higher

This very useful program for observers of Jupiter calculates the positions of Jupiter and its satellites and displays side-on and plan views along with the longitude of Jupiter's central meridian for both equatorial (System I) and nonequatorial (System II) zones. It also displays (and can print) monthly satellite tracks and Great Red Spot transit times and allows animation.

1. Run **js_setup.exe** to intall.

LA LUNA v. 2.03 (. . PC/LUNA)

Contact: Renzo Delrosso, Marco Menichelli, Associazione Astrofili Valdinievole, Biblioteca Comunale, Piazza Martini, I-51015 Monsummano Terme, Italy; Email: Renzo Delrosso@p116.f1.n332.z2. fidonet.org
Type: Shareware
Operating System: DOS

This program provides a wide variety of information about the Moon. Explore a database of lunar phenomena, get the latest ephemerides for the Moon, convert among several time and coordinate systems, or look up dates of past and future lunar and solar eclipses and simulate these events on the screen. The program also features a map of the Moon with information about craters, maria, and other features.

1. Create a directory on your hard disk called **LUNA** and copy all files into it.

2. Exit to DOS, change to the **LUNA** subdirectory, and run **moon203a.exe** through **moon203g.exe** to unpack the software. Enter **LUNA** to run the program.

MARS MOON SIMULATION v. 1.0 (. . PC/MARS)
Contact: Basil N. Rowe, 3489 Observatory Pl., Cincinnati, OH 45208
Type: Shareware
Operating System: DOS

Displays a view of Mars and its moons, Phobos and Deimos, as seen through an inverting telescope, as well as the sub-Earth latitude and longitude of Mars and the position angle of its north pole. The position angle, separation, and magnitude of Phobos and Deimos are also displayed. The view can be animated.

1. Create a directory on your hard disk called **MARS** and copy **marsega1.com** into it.

2. Exit to DOS, change to the **MARS** subdirectory, and run **marsega1.com** to unpack the software. Enter **MARSEGA** to run the program.

MarsIcon v. 2.5 (. . PC/MARSICON)
Contact: W. Scott Thoman, 41 Lee Road, Dryden, NY 13053; Email: thoman@law.mail.cornell.edu
Type: Shareware
Operating System: Windows 3.1 or higher

Generates an icon showing the current view of the planet Mars, similar to the EarthSun program. The display also shows the central meridian of Mars at a particular time. The program can be configured for a particular time zone and Daylight Savings Time.

1. Create a directory on your hard disk called **MARSICON** and copy all files into it.

2. Run **marsicon.exe** and consult the file **read.me** for more information on installing and using the software.

Moon Manager (. . PC/MOONMGR)
Contact: David Kirschbaum; Email: kirsch@usasoc.soc.mil
Type: Freeware
Operating System: DOS

Displays a detailed map of the Moon with a list of craters. By maneuvering the cursor around the map, you can have the program find the nearest major crater and display its name, size, and coordinates. The program can also find a particular feature on the lunar surface or display a list of features in its database.

1. Create a directory on your hard disk called **MOONMGR** and copy all files into it.

2. Exit to DOS, change to the MOONMGR subdirectory, and enter **MOON** to run the program. See **moon.doc** for further information.

MoonTool v. 0.4 (. . PC/MOONTOOL)
Contact: Altmania Productions, Jeffrey Altman, 15 Yarmouth Lane, Nesconset, NY 11767; Email: JALTMAN@ccmail.sunysb.edu
Type: Shareware
Operating System: Windows 3.1 or higher

This program provides updated information on the Moon. For the current time, it computes the Julian date and Universal Time, the phase and age of the Moon, the Moon's distance from the Earth, and the angular diameter the Moon subtends. It also provides the Sun's distance and angular diameter, lists the most recent and upcoming phases of the Moon, and displays a graphic showing the current phase of the Moon.

1. Create a directory on your hard disk called **MOONTOOL** and copy all files into it.

2. Run **moontool.exe** to start the program. See **moontool.txt** for further information.

The Moon Viewer v. 3.01 (. . PC/MOONVIEW)
Contact: Lorenzo Pasqualis; Email: lorenzo@kagi.com
Type: Shareware
Operating System: DOS, Windows 3.1 or higher

Displays the phase of the Moon over wide ranges in time using a photo-realistic view. MoonTool also provides information about the exact phase and distance of the Moon and the times of upcoming new and full Moons and will even generate a printable calendar showing lunar phases during any year you choose.

1. Create a directory on your hard disk called **MOONVIEW** and copy all files into it.

2. Exit to DOS, change to the **MOON** subdirectory, and enter **MOON** to run the program. See **moon.txt** for further information.

MOONBEAM v. 1.0 (. . PC/MOONBEAM)

Contact: Fred Mendenhall, 2209 Tam-O-Shanter Ct., Carmel, IN 46032
Type: Shareware
Operating System: DOS

Displays the Moon relative to background stars for a given date, including its right ascension and declination and its phase. List the minimum star magnitude, and MOONBEAM will plot the position of the Moon and the background stars down to that magnitude.

1. Create a directory on your hard disk called **MOONBEAM** and copy **moonb100.exe** into it.

2. Exit to DOS, change to the **MOONBEAM** subdirectory, and enter **MOONB100** to unpack the software. Enter **MOONBEAM** to run the program.

MOONPHASE v. 2.0 (. . PC/MOONFAZE)

Contact: Scott Baker; Web: 72167.2146@compuserve.com
Type: Freeware
Operating System: DOS

Graphically displays the phase of the Moon for a given time and location. Specify the date, time, latitude, and time zone, and MOONPHASE will compute the rise and set times and the phase of the Moon, as well as the Moon's right ascension and declination. The program will also graphically show the Moon in its current phase to show what features on the disk will be illuminated.

1. Create a directory on your hard disk called **MOONFAZE** and copy **moonfaz.exe** into it.

2. Exit to DOS, change to the **MOONFAZE** subdirectory, enter **MOONFAZE** to extract the software, then enter **MOONFAZE** to run the program.

MoonPhase v. 2.2 (. . PC/MOON22)

Contact: Locutus Codeware; Email: info@locutuscodeware.com; Web: http://www.locutuscodeware.com
Type: Shareware
Operating System: Windows 95 or higher

Displays the current phase of the Moon in the Windows system tray.

1. Run **moonphas.exe** to install.

Moons of Jupiter Orbital Clock v. 1.0 (. . PC/JUPMOONS)

Contact: Andrew Jones, Eagle Rock Village 4-6B, Budd Lake, NJ 07828
Type: Freeware
Operating System: DOS

Simulates the motion of the Galilean moons (Io, Europa, Ganymede, and Callisto) of Jupiter. Enter a date, time, and time zone, and the program displays the locations of the Galilean moons in their orbits around Jupiter as well as their apparent distance from Jupiter as seen from the Earth. The program can also plot the positions of the moons for 3 and 15 days into the future.

1. Create a directory on your hard disk called **JUPMOONS** and copy **jmoons1.exe** into it.

2. Run **JMOONS** to unpack the software then enter **JMOONS** to run the program.

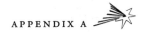

The Nine Planets (. . PC/NINE PLANETS)

Contact: Mp Brown; **Email:** svydog@ix.netcom.com; **Web:** http://
seds.lpl.arizona.edu/nineplanets/nineplanets/intro.html
Type: Freeware
Operating System: Windows 95 or higher

A Windows 95 version of The Nine Planets Web site by Bill Arnett.
Includes detailed information about the Solar System, including
physical data and unanswered questions about each world. Each planetary entry includes several images taken from the Web site. An
overview of the Solar System is also included in the program.

1. Run **setup.exe** to install.

PLANET 1A (. . PC/PLANET)

Contact: Rick Coates
Operating System: DOS

Provides rise and set times, location, and other information about a
specified planet at a given time. Enter all settings for the program at
the command line, including the longitude, latitude, and elevation of an
observer, and the date and time. Specify the rise and set times and the
right ascension and declination coordinates for a particular planet.

1. Create a directory on your hard disk called **PLANET** and copy
   all files into it.

2. Exit to DOS, change to the **PLANET** subdirectory, and enter
   PLANET to run the program. A usage statement on the parameters allowed will appear.

3. Rerun the program and enter the desired parameters on the
   command line before pressing **ENTER**. See **planet.man** for
   additional information.

Planet Guide 96! v. 1.5a (. . PC/PLANGDE)

Contact: Maraj Inc., 861 Pine Ridge Drive, Stone Mnt., GA 30087;
Phone: (770) 413-5838; Fax: (770) 413-5838; Email: maraj@bigfoot.com;
Web: http://www2.gsu.edu/~usgnrmx/maraj/products/prod01.html
Type: Freeware
Operating System: Windows 95 or higher

Provides a brief amount of information on all the planets in our Solar System. Ever wondered what Mars weighs? If so, this handy and informative application can provide an answer for that and numerous other (perhaps less esoteric) questions about the planets circling our Sun.

1. Run **planetg15.exe** and follow the onscreen instructions.

PLANETS v. 4.1 (. . PC/PLANETS)

Contact: Larry Puhl, 6 Plum Ct., Sleepy Hollow, IL 60118
Type: Freeware
Operating System: DOS

Provides information on the location and other attributes of Solar System objects, including what planets are visible in the night sky at a particular time. PLANETS will plot the current locations of the inner and outer planets in their orbits around the Earth and the positions of the moons in their orbits around Mars and the giant outer planets. The date and time and the latitude and longitude of the observer can be changed as well.

1. Create a directory on your hard disk called **PLANETS** and copy the file **planet41.exe** into it.

2. Exit to DOS, change to the **PLANETS** subdirectory, and enter **PLANET41** to unpack the software. Enter **PLANETS** to run the program.

PlanetWatch v. 2.0 (. . PC/PLANWCH)

Contact: Galen Raben, Raben Software & Graphics, P.O. Box 3317, Greeley, CO 80633; Email: galenr@hpgrla.gr.hp.com
Type: Shareware
Operating System: Windows 3.1 or higher

This Solar System reference program includes animated maps, photographs, and other information about the planets in our Solar System in an easy-to-use desktop atlas. Note: Requires the vbrun200.dll and threed.vbx files in the computer's WINDOWS/SYSTEM directory.

1. Create a directory on your hard disk called **PlanetWatch** and copy all files into it. Follow the directions in the **readme.txt** file for installing the software.

THE QUICK ALMANAC v. 1.01 (. . PC/ALMANAC)

Contact: Philip Miller, 32680 Coast Ridge Drive, Carmel, CA 93923
Type: Freeware
Operating System: DOS

Basic information on sunrise and sunset, dawn and dusk, Moon rise and Moon set, the phases of the Moon, and the location (in azimuth and altitude) of the Sun and Moon at a particular time.

1. Create a directory on your hard disk called **ALMANAC** and copy **sun101.com** into it.

2. Exit to DOS, change to the **ALMANAC** subdirectory, and enter **SUN101** to unpack the software, then enter **SUN101** to run the program.

Satellites of Saturn v. 2.0 (. . PC/SATSAT)

Contact: Dan Bruton; Email: astro@tamu.edu
Type: Freeware
Operating System: DOS

Displays the positions of eight of Saturn's largest moons, including Titan, Rhea, Tethys, and Enceladus, at any time. Invert the display to resemble the view through a telescope, or zoom in or out. Orbits can be drawn, and the motions of the moons can be animated. The program can also display a table of current parameters of Saturn and the orbits of the moons.

1. Create a directory on your hard disk called **SATSAT** and copy all files into it.

2. Exit to DOS, change to the **SATSAT** subdirectory, and enter SATSAT2 to run the program.

TRACKER v. 5.0 (. . PC/TRACKER)

Contact: Dan Bruton; Email: astro@tamu.edu
Type: Freeware
Operating System: DOS

Displays the location of Jupiter's Great Red Spot and other features visible on its southern hemisphere, including the locations of the impact sites from the Shoemaker-Levy 9 comet in 1994. Overlay a grid of latitudes and longitudes to help locate objects, animate the map to show the rotation of the planet over time, or display labels with the names of the impact sites and other atmospheric features.

1. Create a directory on your hard disk called **TRACKER** and copy all files into it.

2. Exit to DOS, change to the **TRACKER** subdirectory, and enter **TRACKER5** to run the program. See **tracker5.doc** for further information.

Views of Saturn v. 2 (. . PC/SATVIEW)

Contact: Emanuel Gruengard; Email: f68413@barilvm.bitnet
Type: Freeware
Operating System: DOS

Displays a simulated view of Saturn according to the altitude of the Sun above or below Saturn's ring plane, the altitude of the view point relative to the ring plane, and the longitude difference between the Sun and the view point. The display shows what regions are in sun or shadow, including the region of the rings hidden in the planet's shadow.

1. Create a directory on your hard disk called **SATVIEW** and copy all files into it.

2. Exit to DOS, change to the **SATVIEW** subdirectory, and enter **SATVIEW** to run the program. See **satview.doc** for additional information.

Star Programs

3Star v. 1.1 (. . PC/3STAR)

Contact: Nicholas Law; Email: 3star@dial.pipex.com; Web: http://dialspace.dial.pipex.com/town/road/xpa18/index.shtml
Type: Freeware
Operating System: DOS

View a 3D representation of over 300 stars near the Sun (all positioned accurately according to the Gliese 3.0 catalog). Zoom in or out from 0.5 to 100 light years, or center the display on a particular star.

3Star will also calculate the distance between any two stars, displaying the names and other information about the stars and showing only those stars brighter than a specified magnitude.

1. Create a directory on your hard disk called **3STAR** and copy all files into it.

2. Exit to DOS, change to the **3STAR** subdirectory, and enter **3STAR** to run the program. See **3star.txt** for additional information.

B_STAR (. . PC/BSTAR)

Contact: P.J. Naughter, Cahore, Ballygarrett, Gorey, Co. Wexford, Ireland
Type: Freeware
Operating System: DOS

Computes the observed orientation of a binary star system. The user enters a set of parameters about a binary star system, including the separation of the two stars and the orbit of the secondary star around the primary star as seen from Earth, and for a set of times the program will show where the secondary star would appear relative to the primary star. Enter a range of times to see how the star positions change during orbit—a useful tool for determining how the appearance of a binary system will change over time.

1. Create a directory on your hard disk called **BSTAR** and copy all files into it.

2. Exit to DOS, change to the **BSTAR** subdirectory, and enter **BSTAR** to run the program. Answer the questions about the star system for the program to compute the desired information. See **b_star.doc** for additional information.

BINARY v. 3.0 (. . PC/BINARY)

Contact: Dan Bruton, Texas A&M Physics Department; Email: astro@tamu.edu; Web: http://www.isc.tamu.edu/~astro/
Type: Freeware
Operating System: DOS

Plots the light curve for a binary star—the observed brightness of the binary star as a function of time—which is useful for observing eclipsing binaries (when one star passes in front of another, blocking its light). You can modify the program's star parameters, including the radius and luminosity of each star, the mass ratio of one star to the other, and the orbit of one star around the other.

1. Create a directory on your hard disk called **BINARY** and copy all files into it.

2. Exit to DOS, change to the **BINARY** subdirectory, and enter **BINARY3** to run the program.

3. Modify any of the parameters listed by pressing the number of that parameter. Press **x** to exit. See **binary3.doc** for additional information.

Distant Suns (. . PC/DISSUNS)

Contact: Mike Smithwick; Email: mike@distantsuns.com; Web: http://www.distantsuns.com
Type: Demo (version 4)
Operating System: Windows 3.1 or higher

A good planetarium program that makes effective use of multimedia features such as video. This fully functioning special edition of Distant Suns 4 is missing only the options that require the large database available only by purchasing its CD-ROM (features that include video files, large images, and the 16-million-star database).

1. Run **ds_shareware.exe** to install. See **readme.txt** for additional information.

StarClock v. 2.0 (. . PC/SCLOCK)

Contact: Leos Ondra, Skretova 6, 621 00 Brno, Czech Republic; Email: ondra@sci.muni.cz
Type: Freeware
Operating System: DOS

Animates the evolution of stars on the Hertzprung-Russell (H-R) diagram. Begin with a set of stars of varying solar masses and then watch them evolve from main-sequence hydrogen-burning stars to helium and other phases, starting with the most massive stars. StarClock will also plot the evolution of these stars over time on the H-R diagram. Modify the stars and the time steps used, if you wish.

1. Create a directory on your hard disk called **SCLOCK** and copy all files into it.

2. Exit to DOS, change to the **SCLOCK** subdirectory, and enter **SCLOCK20** to run the program. See **sclock20.doc** for additional information.

Viewers

FITSview (. . PC/FITSVIEW)

Contact: National Radio Astronomy Observatory; Web: http://www. cv.nrao.edu/~bcotton/fitsview.html
Type: Freeware
Operating System: Windows 3.1 (16-BIT) and Windows 95 or higher (32-BIT)

The FITSview family of software consists of viewers for astronomical images in FITS format common to astronomical imaging. These viewers offer a wide variety of image display features and are distributed free of charge by the (U.S.) National Radio Astronomy Observatory. Includes sample FITS images.

1. To install, run **setup.exe** in the CD-ROM directory appropriate to your operating system (see above).

The Image Machine v. 1.0 (. . PC/IMAGE)
Contact: Eddic McCreary; Email: edm@twisto.compaq.com
Type: Freeware
Operating System: Windows 3.1 or higher

Displays images contained on CD-ROMs produced by NASA from Voyager and other mission data in browse file, VICAR, PDS, and BMP formats. Performs a number of operations on the images, including computing their histograms, changing the image palette, and removing reseaus (calibration marks on the images).

1. Create a directory on your hard disk called **IMAGE** and copy all files into it.

2. Exit to DOS, change to the **IMAGE** subdirectory, and enter **Image** to run the program. See **readme.txt** for additional installation instructions.

VuePrint Pro (. . PC/VUEPRINT)
Contact: Hamrick Software; Web: http://www.primenet.com/ ~hamrick
Type: Shareware
Operating System: Windows 95 or higher (vuepro74.exe); Windows 3.1 (vuepri62.exe)

This truly outstanding image viewer and converter for Windows handles an extremely wide range of graphic file formats and also supports many video, animation, and sound formats—and it can act as a scanner interface. Although the trial version we include here times out eventually, this is one application that is well worth registering (and upgrades are free!).

1. Run **vuepro74.exe** (Windows 95 or higher) or **vuepri62.exe** (Windows 3.1) from the CD-ROM.

Macintosh Software

Most of the Macintosh software on the CD-ROM is compressed and must be expanded before it can be installed. If you do not already have StuffIt Expander installed on your system, you should install it before installing any of the software below. (You'll find a copy in the StuffIt directory within the Mac directory on the CD.) Once installed, you should be able to unpack and install the software below.

Galaxy Programs

Galactic Interaction 1.0 (. . Mac/Galactic)
Contact: Neil Schulman; Web: http://www.public.usit.net/nwcs/ index2.html, nwcs@usit.net
Type: Freeware
Operating System: 68020 or higher and System 7.*x* or higher

Simulates the collision of two galaxies and shows how the stars in the target galaxy are affected by the intruder galaxy. Specify the number of rings; the number of stars in each ring of the target galaxy; and the mass, original position, and velocity of the intruder galaxy and watch as the intruder galaxy collides with and scatters the stars of the target galaxy. This is an updated version of the program Galaxy Collisions.

1. Unpack the file **Galactic Interaction.sit.**

2. Double-click the **Galactic Interaction** icon to run the program.

Galaxy Collisions in Three Dimensions (. . Mac/Collisions)
Contact: C. M. Wyatt
Type: Freeware

Simulates the effects on stars when two galaxies collide. Set the type of target galaxy (ring, bridge, or whirlpool), the number of rings of stars in the galaxy, and the number of stars in each ring and choose the mass, original position, and velocity of the intruder galaxy—then have Galaxy Collisions simulate and display the collision of the two galaxies. Simulations can be saved or printed.

1. Run the file **3dgc.sea** to unpack the software.

2. Double-click the icon **Galaxy Collisions** to start the program. See **Help Text 3DGC** for help.

Gravity Programs

Gravitation Ltd. 4.0 (. . Mac/Gravitation)
Contact: Jeff Rommereide; Email: J.ROMMEREIDE@genie.geis.com
Type: Shareware

Simulates the effects of gravity on a set of objects. Choose the number of bodies, their masses, their original positions, and locations; then simulate the effects of the gravity of each object on every other object. Change the speed of the simulation or draw trails marking the paths followed by the objects.

1. Run the file **gravit40.sea** to unpack the software.

2. Double-click the icon labeled **Gravitation Ltd. 4.0** to run the program.

Multipurpose Programs

SkyChart III (. . Mac/SkyChart)
Contact: Southern Stars Software; Web: http://www.southernstars. com/skychart/home.html; Email: info@southernstars.com
Type: Demo
Operating System: MacOS System 7 or higher; 68020 or higher, or PowerPC

This excellent and reasonably priced astronomy application includes a wealth of features and is available for a wider variety of computing platforms than similar programs. In addition to providing an excellent planetarium function, SkyChart III can compute and display the positions of Earth satellites using standard NASA/NORAD two-line element satellite orbit files and print high-resolution star charts and highly detailed ephemerides. It also includes a fully customizable object database containing more than 40,000 stars and hundreds of Messier, NGC, and IC deep-sky objects. All existing objects can be edited, including the constellations, and you can import your own database to add new comets, asteroids, stars, planets, or deep-sky objects.

1. To install, execute **SkyChart_III_Demo_Installer.bin**.

Utility Programs

Adobe Acrobat Reader (. . Mac/Adobe)
Contact: Adobe Systems; Web: http://www.adobe.com
Type: Freeware
Operating System: 68020 or later; Power PC with Apple System
Software 7.0 or later (7.12 or later recommended).

Several applications on this CD-ROM include documents in the Adobe
Acrobat Portable Document Format (PDF), so we're including the free
Adobe Acrobat Reader software so users can view these documents.

1. Double-click **Install Acrobat Reader 4.0.** (Note: This will
   cause all current Macintosh applications to quit); then follow
   the instructions on your screen.

QuickTime 4 (. . Mac/QuickTime)
Contact: Apple Computer; Web: http://www.apple.com/quicktime
Type: Freeware
Operating System: Mac OS 7.1 or later; 68020 or later; Power Macs.
QuickTime 4 is freeware, but registering the application turns it into
the Pro version and enables many extra features.

QuickTime is an Apple technology that lets you view and create
video, sound, music, 3D, and virtual reality for both Macintosh and
Windows systems. Several of the animations and panoramas found
on this CD-ROM require QuickTime for viewing.

1. Run **QuickTimeInstaller.bin** then run **QuickTime Installer.**

Stuffit Expander 5.1.2 (. . Mac/Stuffit)
Contact: Aladdin Systems; **Web:** http://www.aladdinsys.com
Type: Freeware
Operating System: System 7.1.1 or higher and 68020 or better

Use Stuffit Expander to decompress or decode files. Just drag-and-drop a compressed or encoded file on Stuffit Expander and the application accesses the file automatically.

1. Run **aladdin_exp_512_installer.bin** then run **Aladdin Expander 5.1.2 Install.**

Observational Programs

Astroeve 4.2 (. . Mac/Astroeve)
Type: Freeware

Displays the next 12 times of a set of astronomical events. Calculates the times of upcoming seasonal solstices and equinoxes; full Moons; inferior conjunctions of Mercury and Venus; and oppositions of Mars, Jupiter, and Saturn. Operated by a simple set of buttons.

1. Double-click the icon for **astroeve.bin** to run the program.

Mac Ephem (. . Mac/Ephem)
Contact: Peter Newton
Type: Freeware

Computes the observation circumstances for a particular time. For each planet, along with the Sun and Moon, the program calculates and displays in a text table the current position, aziumth, and latitude; distance from the Sun and Earth; angular size; brightness; and phase. The program also displays the current Julian date, Universal Time, time of dawn and dusk, and upcoming new and full Moons.

1. Run the file **mephem421.sea** to unpack the software; then double-click the **mephem4.21** icon to run the program. See **Ephem4.20.man** for help.

Star Atlas (. . Mac/Star atlas)

Contact: Youhei Morita, Computing Research Center, High Energy Accelerator Research Organization (KEK), 1-1 Oho, Tsukuba, Ibaraki 305-0801, Japan; Phone: +81-298-64-5476; Fax: +81-298-64-4402; Email: Youhei.Morita@kek.jp
Type: Freeware
Operating System: 68k Mac or higher; System 7.1 or higher.

Displays a full-screen view of the night sky, with over 1,500 stars available. The map can be centered or zoomed in on a particular location using the menus or by clicking on a location with the mouse. Star names can be displayed for brighter stars.

1. Run the file **satl0b1.sea** to unpack the software; then double-click the **Star Atlas** icon to run the program.

Sunrise (. . Mac/Sunrise)

Contact: Leighton Paul
Type: Shareware

Calculates and displays the times of sunrise and sunset, as well as the times of civil, nautical, and astronomical twilight.

1. Double-click the icon **sunris11.bin** to run the program.

Telescopes 1.02 (. . Mac/Telescopes)

Contact: Brian D. Monson, Email: bmons@ctechok.org
Type: Freeware **Requirements:** Hypercard 2.1, QuickTime 1.5

This Hypercard stack introduces various types of telescopes and instru- ments used on these telescopes. It covers topics such as light-gathering ability, resolving power, and magnification of telescopes; various telescope designs (including ray tracings to show the path light takes through the telescopes); spectroscopes and CCD cameras; and buying tips. It requires the Hypercard Reader (freely available and found on most Macintosh systems).

1. Use Stuffit Expander to unpack the file **telescopes102.sit**; then double-click the **Telescopes 1.02** icon to run the stack. See **Scopes Documentation** for futher information.

Planetarium Programs

MacAstro 1.6 (. . Mac/MacAstro)

Contact: Nicolas Mercouroff, mercouroff@aar.alcatel-alsthom.fr
Type: Shareware **Operating System:** Macintosh Plus, Classic, or higher; System 6.0 or higher.

This simple planetarium program displays the positions of over 2,500 stars at a given time and location as well as the current sidereal time, the locations of the planets, the phase of the Moon, and the positions of the satellites of Jupiter and Saturn. Clicking on an object displays more information about it, including its name, coordinates, and rise and set times for a specified location. Select stars from a menu to show the same information and to highlight them on the map.

1. Use Stuffit Expander to unpack **macastro-1.6.sit**; then double-click the icon **MacAstro 1.6** to run the program. See Doc **MacAstro 1.6** for further information.

NightSky 2.21 (. . Mac/NightSky)

Contact: Kaweah Software; Web: http://www.kaweah.com/night-sky.html; Email: djensen@kaweah.com

Type: Freeware **Operating System:** Power Macintosh or older 680x0 Macs with an FPU (Floating Point Unit); most 68020/30/40 Macs running Software FPU.

This simple planetarium program displays the night sky with up to 50,000 stars as well as the Sun, Moon, and planets. Set the limiting magnitude of the display (between magnitudes 2 and 8), draw constellation lines and star names, and set the display type to one of four projections (orthographic, Mercator, polar, and plate carie) at any given time. You can even draw on the display. Export images to PICT or PNG format.

1. Unpack the file **NighySky-2.21.sit**; then double-click the **NightSky** icon to run the program.

Starry Night Basic 2.1.3 (. . Mac/Starry Night)

Contact: Sienna Software, Inc; Web: http://www.siennasoft.com/; Email: support@siennasoft.com

Type: Shareware **Operating System:** System 7.0 or higher

View the night sky from any place on Earth or in the Solar System at any time. A "Swiss Army Knife" Astronomy application, Starry Night is very feature rich and is being used around the world by parents, educators, amateur astronomers, and those just beginning to discover the wonders of astronomy. The deluxe CD-ROM version adds many features, including a greatly expanded database of stars and of comets whose tails grow in size as they approach the Sun; the Macintosh version offers a plug-in that can control computerized telescopes.

1. Run the self-extracting archive **starrynightbasic_sit.hqx** to unpack the software then run **Starry Night Basic Installer** to install.

Satellite Programs

OrbiTrack (. . Mac/OrbiTrack)
Contact: BEK Developers, PO Box 47114, St. Petersburg, FL 33743-7114; Email: 75366.2557@compuserve.com
Type: Shareware

Tracks spacecraft orbiting the Earth. Track the position of satellites, plot their ground tracks on a map of the Earth for a specified time, do real-time tracking, show the passage of satellites against a map of background stars, and calculate Sun rise and set times. Read in satellite and ground station data or provide your own, choose from among several orbit models, and write satellite data to a file or copy it to the clipboard.

1. Unpack **OrbiTrack.sit**; then double-click the icon **OrbiTrack 2.1.5** to run the program.

SatTrak 1.02 (. . Mac/SatTrak)
Contact: Mike Pfleuger; Web: http://www.primenet.com/~pflueg/; Email: pflueg@primenet.com
Type: Shareware
Operating System: System 6.0 or higher

This program is designed to help amateur radio operators track Earth-orbiting satellites for use in communications. For a specified satellite and location, the program will compute when the satellite is above the horizon, including its azimuth and altitude angles, distance and altitude, and the Doppler shift for any radio waves coming from the satellite. Also performs calculations for radio communications and calculates the antenna types and sizes needed for specified wavelengths.

1. Run the file **SatTrak.sea** to unpack the software, then double-click the **SatTrak** icon to start the program.

Solar System Programs

EarthPlot (. . Mac/EarthPlot)
Contact: The Black Swamp Software Company
Type: Freeware

Generates globes of the Earth. Center the globe on a particular longitude and latitude and determine the altitude from which it should be viewed (from 1,000 km to 160,000 miles), and EarthPlot draws a line diagram of the selected view of the Earth, including longitude and latitude lines. Copy and paste the diagram into other graphics or drawing programs, as desired.

1. Run the file **eplot30.sea** to unpack the software; then double-click the file **Earthplot 3.0** to run the software. See **Using Earth Plot** for further information.

JSAT (. . Mac/Jsat)
Contact: Mike Kazmierczak, X Systems, 1789 Brandy Dr. SE, Conyers, GA 30208
Type: Shareware

Displays the locations of the four Galilean satellites of Jupiter at a given time in various formats, including arc seconds; graphical depictions of Jupiter and the moons; and a trace of the moons' distances from Jupiter as a function of time.

1. Run the file **jsat1.sea** to unpack the software; then double-click the file JSAT to run the program. See **jupiter.doc** for further information.

Jupiterium (. . Mac/Jupiterium)
Contact: Howard L. Ritter, Jr., The Eccentric Anomaly, 9220 Jarwoods Dr., Texas 78250
Type: Freeware

Graphically displays the locations of the Galilean satellites of Jupiter at a given time. Simulate the view through several types of telescopes, including those that invert or reverse the image, like Newtonian and Schmidt-Cassegranian telescopes. Set the time and time zone and animate the view of the moons.

1. Run the file **jupiteri.sea** to unpack the software; then double-click the icon for **Jupiterium v2.1** to run the program.

MacSATSAT (. . Mac/MacSatsat)
Contact: Dan Bruton & Neil Simmons; Email: Neil.Simmons@m.cc.utah.edu or Dan Bruton (astro@tamu.edu)
Type: Freeware

Displays the location of Saturn's major moons—Mimas, Enceladus, Tethys, Dione, Rhea, Titan, Hyperion, and Iapetus—from two orientations. The view can be zoomed in and out, and the orbits of the moons can be drawn on the screen.

1. Double-click the icon **MacSATSAT** to run the program.

Moon (. . Mac/Moon)
Contact: David Palmer, palmer@caltech.edu
Type: Freeware

Displays a black-and-white graphic showing the current phase of the Moon at a particular time.

1. Run the file **moon.sea** to unpack the software; then double-click the moon icon to run the program.

MoonPhaser 1.01 (. . Mac/MoonPhaser)

Contact: Dave Kalin

Shows the phase of the Moon. The default is the current phase of the Moon, but MoonPhaser will also show phases of the Moon on the previous and next days, search for the dates of the next new and full Moons, or check the phase for any date from 1984 through 2006.

1. Run the file **moonph10.sea** to unpack the software; then double-click the Moonphaser icon to start the program. See **moon.doc** for further information.

Orrery 1.0 (. . Mac/Orrery)

Contact: John de Boer, 60-A Lundy's Lane, Kingston, Ontario Canada K7K5G7
Type: Shareware
Operating Sytem: Power PC, System 7.0

Simulates an orrery, a device that shows the motions of planets, the Moon, and asteroids in our Solar System. Select the bodies to observe and watch them orbit forward or backward in time at various speeds. Lock onto a particular planet and watch its motion around the Sun; watch satellites in orbit around the Earth; and view ray-traced maps of planetary surfaces drawn onto each planet in the orrery. Images can be saved as PICT image files for later use.

1. Unpack the file **orrery.sit** then double-click the **Orrery** icon to start the program. See **Instructions** for additional information.

Shadow (. . Mac/Shadow)

Contact: Steve Splonskowski, Email: splons@teleport.com
Type: Freeware

Shows where the Sun is visible from the Earth at the current time using a world map displaying regions of the Earth in sun and shadow.

1. Use Stuffit Expander to unpack the file **Shadow.sit**.

2. Double-click the **Shadow** icon to run the program.

Xearth for Mac 1.02 (. . Mac/Xearth)

Contact: Johan Lindvall; Web: http://www.dtek.chalmers.se/~d2linjo/ mac_xearth.html; Email: d2linjo@dtek.chalmers.se
Type: Freeware
Operating System: System 7.0 or higher

Displays a color view of the Earth according to your specifications (whether the view should be fixed on the sunlit side of the Earth or fixed on a specific location). Display Earth cities or stars in the background. The view can be fixed at a specific time, set to update to the current time, and resized.

1. Unpack the file **Xearth for Mac.sit**; then double-click the **Xearth for Mac** icon to run the program.

Star Programs

Distant Suns (. . Mac/Distant Suns)

Contact: Mike Smithwick; Web: http://www.scruz.net/~gosborn/distant/index.html; Email: gosborn@scruznet.com
Type: Demo
Operating System: System: 7.1 or higher (8.0 recommended); PowerPC 601 or higher; 9MB RAM; 15MB hard disk space; Quicktime

A good planetarium program that makes effective use of multi-media features such as video. This fully functioning special edition of Distant Suns 4 is missing only the options that require the large database available only by purchasing its CD-ROM (features that include video files, large images, and the 16-million-star database).

1. Run the file **ds_installer. bin**. Documentation is contained in DS5_0MacUserGuide.pdf.

Sol's Neighbors (. . Mac/Sol)
Contact: John Cramer; Web: http://weber.u.washington.edu/~jcramer/stacks.html; Email: cramer@phys.washington.edu
Type: Freeware
Requirements: Hypercard

A Hypercard stack with information about the stars nearest the Sun. Lists a wide range of information for each star, including its location, parallax, distance and proper motion, spectral class, luminosity, surface temperature, magnitude, and mass and radius relative to the Sun. Can also plot the position of the star relative to the Sun in three dimensions. Requires the Hypercard Reader (freely available and found on most Macintosh systems).

1. Run the file **solsnghb.sea** to unpack the software; then double-click the icon **sol'sNeighbors** to run the stack.

Viewers

MacFITSview (. . Mac/Fitsview)
Type: Freeware
Operating System: Macintosh System 7.0, 68000 series, Power Mac

The FITSview family of software consists of viewers for astronomical images in FITS format common to astronomical imaging.

These viewers offer a wide variety of image display features and are distributed free of charge by the (U.S.) National Radio Astronomy Observatory.

1. Run **MacFITSView_68k.hqx** for 68000 sytems or **MacFITSView_PPC.hqx** for Power PCs.

APPENDIX B

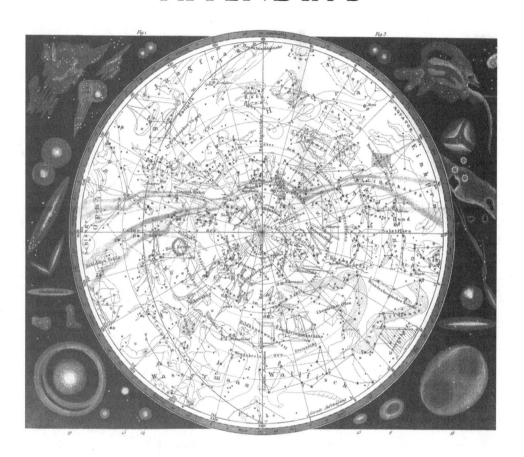

IMAGES ON THE CD-ROM

The collection of images included here was selected based on several different criteria. First, they had to succeed as good photographs—visually interesting, well exposed and composed. Secondly, they were chosen for their diversity—they needed to cover a broad spectrum of objects that ranged from the Earth's Grand Canyon to the faintest galaxies recorded by long exposures using the Hubble Space Telescope. Finally, we chose the highest resolution images we were able to locate for any particular object thus allowing you to "zoom in" on many of these images and see still more detail.

Images were selected up until the last possible moment before the CD-ROM was submitted. More images are released somewhere almost daily, so browsing the Internet for new astronomical images can be a very rewarding experience.

Note: All images are listed in alphabetical order.

26301_02_olympus.jpg

A wide-angle view of the Martian volcano Olympus Mons on 25 April 1998, taken by the Mars Global Surveyor (MGS) Mars Orbiter Camera (MOC). This is a false color image; the actual colors would be paler and the contrast more subdued. North is to the left in this image, east is toward the top; the image was taken 560 miles from the Martian surface. Olympus Mons is taller than Mount Everest and wider than the Hawaiian Island chain.

408_allsky_big.gif

The sky at a radio frequency of 408MHz, based on observations of the radio observatories Jodrell Bak, MPIfR and Parkes Observatory. This false color image has the galactic plane running horizontally through the center, with the galaxy Centaurus A just above the plane at the right of center and the Large Magellanic Cloud below and to the right. Image credit: C. Haslam et al., MPIfR, SkyView.

550_1_2_proj.jpg

The north polar region of Mars as photographed by the Mars Global Surveyor's (MGS) Mars Orbital Camera (MOC). This is a false color image made by creating red and blue filtered images and synthesizing the green layer by averaging the red and blue blands. The images have been reprojected to simulate the view a person would have from about 740 miles Mars at 65°N, 275° W. The swirled pattern at the top is caused by layered deposits that are partly covered by the permanent north polar ice cap.

96-09a.jpg

A Hubble Space Telescope image of the planet Pluto, taken with the European Space Agency's Faint Object Camera (FOC). This is STScI-PRC96-09a, credited to A. Stern (SwRI), M. Buie (Lowell Obs.), NASA, and ESA.

96-23a.jpg

A pair of gas and dust clouds in the Eta Carinae nebula, as photographed by the Hubble Space Telescope. This is STScI-PRC96-23a, credited to Jon Morse (University of Colorado) and NASA.

96-38a.jpg

Giant tornado-like structures in the Lagoon Nebula (M8), as photographed by the Hubble Space Telescope. Each of the structures is about one-half light-year long. The Lagoon Nebula is in the constellation Sagittarius, and is about 5,000 light-years from Earth. This is STScI-PRC96-38a, credited to NASA.

97-11.jpg

The Egg Nebula (CRL 2688) is shown on the left as it appears in visible light to the Hubble Space Telescope, and on the right as it appears in infrared light to Hubble's Near Infrared Camera and Multi-Object Spectrometer (NICMOS). The infrared image has been assigned colors to distinguish different wavelengths, with blue corresponding to starlight reflected by dust particles and red corresponding to heat radiation emitted by hot molecular hydrogen.

This is STScI-PRC97-11, and is attributed to Rodger Thompson, Marcia Rieke, Glenn Schneider, Dean Hines (University of Arizona); Raghvendra Sahai (Jet Propulsion Laboratory); NICMOS Instrument Definition Team, and NASA.

9734ay.jpg

A Hubble Space Telescope image showing the center of a collision between two galaxies. This is STScI-PRC97-34a, credited to Brad Whitmore (STScI) and NASA.

9738ay.jpg

M2-9 is a bipolar (butterfly) planetary nebula about 2,100 light years from Earth in the constellation Ophiucus, shown here in an image from the Hubble Space Telescope's Wide Field and Planetary Camera 2. In this image, neutral oxygen is shown in red, once-ionized nitrogen in green, and twice-ionized oxygen in blue. This is STScI-PRC97-38, credited to Bruce Balick (University of Washington), Vincent Icke (Leiden University, The Netherlands), Garrelt Mellema (Stockholm University), and NASA.

9738by.jpg

A Hubble Space Telescope gallery of planetary nebulae:

(Top left) IC 3568 lies in the constellation Camelopardalis at a distance of about 9,000 light-years, and has a diameter of about 0.4 light-years (or about 800 times the diameter of our Solar System). It is an example of a round planetary nebula. Note the bright inner shell and fainter, smooth, circular outer envelope. Credits: Howard Bond (Space Telescope Science Institute), Robin Ciardullo (Pennsylvania State University), and NASA.

(Top center) NGC 6826's eye-like appearance is marred by two sets of blood-red "fliers" that lie horizontally across the image. The surrounding faint green "white" of the eye is believed to be gas that made up almost half of the star's mass for most of its life. The hot remnant star (in the center of the green oval) drives a fast wind into older material, forming a hot interior bubble that pushes the older gas ahead of it to form a bright rim. (The star is one of the brightest

stars in any planetary.) NGC 6826 is 2,200 light-years away in the constellation Cygnus. The Hubble telescope observation was taken January 27, 1996 with the Wide Field and Planetary Camera 2. Credits: Bruce Balick (University of Washington), Jason Alexander (University of Washington), Arsen Hajian (U.S. Naval Observatory), Yervant Terzian (Cornell University), Mario Perinotto (University of Florence, Italy), Patrizio Patriarchi (Arcetri Observatory, Italy), and NASA.

(Top right) NGC 3918 is in the constellation Centaurus and is about 3,000 light-years from us. Its diameter is about 0.3 light-year. It shows a roughly spherical outer envelope but an elongated inner balloon inflated by a fast wind from the hot central star, which is starting to break out of the spherical envelope at the top and bottom of the image. Credits: Howard Bond (Space Telescope Science Institute), Robin Ciardullo (Pennsylvania State University), and NASA.

(Bottom left) Hubble 5 is a striking example of a "butterfly" or bipolar (two-lobed) nebula. The heat generated by fast winds causes each of the lobes to expand, much like a pair of balloons with internal heaters. The Hubble telescope's Wide Field and Planetary Camera 2 took this observation September 9, 1997. Hubble 5 is 2,200 light-years away in the constellation Sagittarius. Credits: Bruce Balick (University of Washington), Vincent Icke (Leiden University, The Netherlands), Garrelt Mellema (Stockholm University), and NASA.

(Bottom center) Like NGC 6826, NGC 7009 has a bright central star at the center of a dark cavity bounded by a football-shaped rim of dense blue and red gas. The cavity and its rim are trapped inside smoothly distributed greenish material in the shape of a barrel and comprised of the star's former outer layers. At larger distances, and lying along the long axis of the nebula, a pair of red "ansea", or "handles," appears. Each ansa is joined to the tips of the cavity by a long greenish jet of material. The handles are clouds of low-density gas. NGC 7009 is 1,400 light-years away in the constellation Aquarius. The Wide Field and Planetary Camera 2 took the Hubble telescope observation April 28, 1996. Credits: Bruce Balick (University of Washington), Jason Alexander (University of Washington), Arsen Hajian (U.S. Naval Observatory), Yervant Terzian (Cornell University), Mario

Perinotto (University of Florence, Italy), Patrizio Patriarchi (Arcetri Observatory, Italy), and NASA.

(Bottom right) NGC 5307 also lies in Centaurus but is about 10,000 light-years away and has a diameter of approximately 0.6 light-year. It is an example of a planetary nebula with a pinwheel or spiral structure; each blob of gas ejected from the central star has a counterpart on the opposite side of the star. Credits: Howard Bond (Space Telescope Science Institute), Robin Ciardullo (Pennsylvania State University), and NASA.

9804y.jpg

A complete view of Jupiter's northern and southern auroras, taken in ultraviolet light by the Hubble Space Telescope's Space Telescope Imaging Spectrograph (STIS). Both auroras show vapor trails of light left by Jupiter's moon Io; the vapor trails are the white, comet-shaped streaks just outside both auroral ovals. These streaks are not part of the auroral ovals, but are caused when an electrical current of charged particles are ejected from Io and flow along Jupiter's magnetic field lines to the planets north and south magnetic poles. This electrical charge equals about 1 million amperes. This photograph was taken on September 20, 1997, and is shown here in false color. The planet's reflected sunlight appears brown in this image, while the auroral emissions appear white or shades of blue and red. Photo credit: John Clarke (University of Michigan) and NASA.

9811ay.jpg

Two images of the planetary nebula NGC 7027 made by the Hubble Space Telescope; on the left is an composite infrared image, on the right is a composite image in both visible and infrared light. In the image on the right, white indicates emission from the hot gas surrounding the central star, and pink and red represent emissions from cool molecular hydrogen as. NGC 7027 is in the constellation Cygnus, and is located about 3,000 light-years distant. Photo credit: William B. Latter (SIRTF Science Center/Caltech) and NASA.

9825y.jpg

This "nearby" massive star cluster, N81, is located some 200,000 light-years away in the Small Magellanic Cloud and was imaged by the Hubble Space Telescope. The Hubble image shows some fifty separate stars packed in a core only 10 light-years in diameter. The accompanying nebula is sculpted by violent stellar winds and shockwaves by the phenomenal mass loss from these super-hot stars, each with the brilliance 300,000 times that of our Sun.

This "natural-color" view was assembled from separate images taken with the Wide Field and Planetary Camera 2, in ultraviolet light and two narrow emission lines of ionized Hydrogen (H-alpha, H-beta). The image is credited to Mohammad Heydari-Malayeri (Paris Observatory, France), and NASA/ESA.

9832ay.jpg

These images represent the Hubble Space Telescope's deepest view into the universe to date. The image on the left is taken in infrared light and contains over 300 galaxies, some of which may be over 12 billion light-years distant. The "field of view" of this image is 2 million light-years across at its maximum, but represents an area of the sky one-thousandth the apparent diameter of the full moon. The two close-up images on the right are views of objects suspected to be over 12 billion light-years away. The image was taken in January 1998 and required an exposure time of 36 hours to detect images down to 30th magnitude. The area of the sky imaged is just above the handle of the Big Dipper, in the constellation Ursa Major. This is STScI-PRC98-32, and is credited to Rodger L. Thompson (University of Arizona), and NASA.

9838y.jpg

A Wolf-Rayet star, WR124, surrounded by clumps of hot gas that are being ejected into space at over 100,000 miles per hour. This false color image was taken with the Hubble Space Telescope's Wide Field Planetary Camera 2, and shows the nebula M1-67 surrounding WR124. WR124 is 15,000 years distant, in the constellation Sagittarius. Credit for this image goes to Yves Grosdidier (University of Montreal and Observatoire de Strasbourg), Anthony Moffat (Universitie de Montreal), Gilles Joncas (Universite Laval), Agnes Acker (Observatoire de Strasbourg), and NASA.

9839a.jpg

NGC 3132, The Eight-Burst Nebula, as photographed by the Hubble Space Telescope. This planetary nebula is located in the constellation Vela, and is about 0.4 light-year in diameter and some 2,000 light-years distant. Two stars are obvious near the center of NGC 3132—the fainter star is the star that has ejected the surrounding nebula. This is a false color image in which blue represents the hottest gas, which is confined to the inner region of the nebula, and red represents the coolest gas at the outer edge.

Credit for this image goes to the Hubble Heritage Team (AURA/ STScI/NASA); principal astronomers were Raghvendra Sahai, John Trauger, and Robin Evans (Jet Propulsion Lab), and the WFPC2 Investigation Definition Team (IDT).

9842w.jpg

NGC 253 is a large, almost edge-on spiral galaxy, and is one of the nearest galaxies beyond our local neighborhood of galaxies. This dramatic galaxy shows complex structures such as clumpy gas clouds, darkened dust lanes, and young, luminous central star clusters. These elements are typical of spiral galaxies. Caroline Herschel discovered NGC 253 in 1783 while looking for comets. The galaxy's closeness to Earth makes it an ideal target for amateur astronomers who can see the southern sky and for astronomers interested in learning more about the makeup of these stunning cities of stars. Photographed with the Hubble Space Telescope. Credit: Hubble Heritage Team (AURA/STScI/NASA).

9901a.jpg m57ring_hst_big.jpg (larger version)

M57 (NGC 6720), the famous Ring nebula in the constellation Lyra, as photographed by the Hubble Space Telescope. The planetary nebula is approximately a light-year in diameter and about 2,300 light-years distant. In this image, the colors are generally realistic; they are based on three chemical elements: helium (blue), oxygen (green), and nitrogen (red). This Hubble image required an exposure time of one hour. Credit for this image goes to the Hubble Heritage Team (AURA/STScI/NASA). Principal astronomers were H. Bond, C. Christian, J. English, L. Frattare, F. Hamilton, A. Kinney, Z. Levay, and K. Noll.

9905ay.jpg

The following images were taken by NASA Hubble Space Telescope's Near-Infrared Camera and Multi-Object Spectrometer (NICMOS). All of the objects are extremely young stars, 450 light-years away in the constellation Taurus. Most of the nebulae represent small dust particles around the stars, which are seen because they are reflecting starlight. In the color-coding, regions of greatest dust concentration appear red. All photo credits: D. Padgett (IPAC/Caltech), W. Brandner (IPAC), K. Stapelfeldt (JPL), and NASA.

(Top left) CoKu Tau/1. This image shows a newborn binary star system, CoKu Tau/1, lying at the center of four "wings" of light extending as much as 75 billion miles from the pair. The "wings" outline the edges of a region in the stars' dusty surroundings, which have been cleared by outflowing gas. A thin, dark lane extends to the left and to right of the binary, suggesting that a disk or ring of dusty material encircles the two young stars.

(Top center) DG Tau B. An excellent example of the complementary nature of Hubble's instruments may be found by comparing the infrared NICMOS image of DG Tau B to the visible-light Wide Field and Planetary Camera 2 (WFPC2) image of the same object. WFPC2 highlights the jet emerging from the system, while NICMOS penetrates some of the dust near the star to more clearly outline the 50 billion-mile-long dust lane (the horizontal dark band, which indicates the presence of a large disk forming around the infant star). The young star itself appears as the bright red spot at the corner of the V-shaped nebula.

(Top right) Haro 6-5B. This image of the young star Haro 6-5B shows two bright regions separated by a dark lane. As seen in the WFPC2 image of the same object, the bright regions represent starlight reflecting from the upper and lower surfaces of the disk,

which is thicker at its edges than its center. However, the infrared view reveals the young star just above the dust lane.

(Bottom left) I04016. A very young star still deep within the dusty cocoon from which it formed is shown in this image of IRAS 04016+2610. The star is visible as a bright reddish spot at the base of a bowl-shaped nebula about 100 billion miles across at the widest point. The nebula arises from dusty material falling onto a forming circumstellar disk, seen as a partial dark band to the left of the star. The necklace of bright spots above the star is an image artifact.

(Bottom center) I04248. In this image of IRAS 04248+2612, the infrared eyes of NICMOS peer through a dusty cloud to reveal a double-star system in formation. A nebula extends at least 65 billion miles in opposite directions from the twin stars and is illuminated by them. This nebula was formed from material ejected by the young star system. The apparent "pinching" of this nebula close to the binary suggests that a ring or disk of dust and gas surrounds the two stars.

(Bottom right) I04302. This image shows IRAS 04302+2247, a star hidden from direct view and seen only by the nebula it illuminates. Dividing the nebula in two is a dense, edge-on disk of dust and gas which appears as the thick, dark band crossing the center of the image. The disk has a diameter of 80 billion miles (15 times the diameter of Neptune's orbit), and has a mass comparable to the Solar nebula, which gave birth to our planetary system. Dark clouds and bright wisps above and below the disk suggest that it is still building up from infalling dust and gas.

9905by.jpg

Hubble Sees Disks Around Young Stars: Credits: Chris Burrows (STScI), John Krist (STScI), Karl Stapelfeldt (JPL) and colleagues, the WFPC2 Science Team, and NASA.

(Top left) This Wide Field and Planetary Camera 2 (WFPC2) image shows Herbig-Haro 30 (HH 30), the prototype of a young star surrounded by a thin, dark disk and emitting powerful gaseous jets. The disk extends 40 billion miles from left to right in the image, dividing the nebula in two. The central star is hidden from direct view, but its light reflects off the upper and lower surfaces of the disk

to produce the pair of reddish nebulae. The gas jets are shown in green. Credit: Chris Burrows (STScI), the WFPC2 Science Team, and NASA.

(Top right) DG Tauri B appears very similar to HH 30, with jets and a central dark lane with reflected starlight at its edges. In this WFPC2 image, the dust lane is much thicker than seen in HH 30, indicating that dusty material is still in the process of falling onto the hidden star and disk. The bright jet extends a distance of 90 billion miles away from the system. Credit: Chris Burrows (STScI), the WFPC2 Science Team, and NASA.

(Lower left) Haro 6-5B is a nearly edge-on disk surrounded by a complex mixture of wispy clouds of dust and gas. In this WFPC2 image, the central star is partially hidden by the disk, but can be pinpointed by the stubby jet (shown in green), which it emits. The dark disk extends 32 billion miles across at a 90-degree angle to the jet. Credit: John Krist (STScI), the WFPC2 Science Team, and NASA.

(Lower right) HK Tauri is the first example of a young binary star system with an edge-on disk around one member of the pair. The thin, dark disk is illuminated by the light of its hidden central star. The absence of jets indicates that the star is not actively accreting material from this disk. The disk diameter is 20 billion miles. The brighter primary star appears at top of the image. Credit: Karl Stapelfeldt (JPL) and colleagues, and NASA.

9908a.jpg

On April 6, 1994, NASA's Hubble Space Telescope (HST) was performing a detailed study of the Sun's nearest stellar neighbor, Proxima Centauri, using the Fine Guidance Sensors to search for small deviations in the position of Proxima Centauri that could reveal the presence of an unseen planetary companion. Rather than sit idle while this study went on, the Wide Field and Planetary Camera 2 (WFPC2) was activated using the observing strategy set out in a program initiated by Dr. Ed Groth (Princeton) designed to make use of this otherwise wasted time. The image captured by this WFPC2 parallel observation is a typical Milky Way star field in the constellation Centaurus. Such images can be used to study the evolution of stars that make up our galaxy. Most of the stars in this image lie near the center of our galaxy some 25,000 light-years distant. But one object, the blue curved streak, is something much closer. An uncatalogued, mile-wide bit of rocky debris orbiting the Sun only light-minutes away strayed into WFPC2's field while the image was being exposed. This and about a hundred other interlopers have been found by Jet Propulsion Laboratory astronomers Dr. Robin Evans, Dr. Karl Stapelfeldt, and collaborators, who have systematically searched the HST archive for these nearby objects. Their analysis indicates this asteroid's orbit could cross Mars' path. Seen briefly by HST, these asteroids are too small and faint to track from the ground long enough for precise orbits to be determined. They are destined to return to their unseen wanderings for hundreds or thousands of years until once again, by chance, they may flicker across the view of some watchful eye peering off into the depths of space.

This particular image is a 33-minute exposure. Principal astronomers: Edward J. Groth (Princeton) and collaborators. Image and commentary credit: Hubble Heritage Team (AURA/STScI/NASA). Asteroid Designation: u2805m01t4, Magnitude: 18.7, Diameter: 1.25 miles (2 kilometers); the asteroid is 87 million miles (0.93 A.U.) from Earth and 156 million miles (1.68 A.U.) from the Sun.

9912a.jpg

Hodge 301, a star cluster inside the Tarantula Nebula found in the Large Magellanic Cloud, appears at the lower right of this Hubble Space Telescope image. Near the center of the image are small gas globules and dust columns where new stars are being formed. Hodge 301 is located in the constellation Dorado, and is some 168,000 light-years distant. This image required an exposure time of 1.8 hours, and is roughly 96 light years on the vertical side. Principal astronomers were You-Hua Chu (University of Illinois), Eva K. Grebel (University of Washington), and collaborators. Image credit: Hubble Heritage Team (AURA/STScI/NASA).

9916a.jpg

Located about 130 million light-years away, NGC 4650A is one of only 100 known polar-ring galaxies. Their unusual disk-ring structure is not yet understood fully. One possibility is that polar rings are the remnants of colossal collisions between two galaxies sometime in the distant past, probably at least 1 billion years ago. During the collision the gas from a smaller galaxy would have been stripped off and captured by a larger galaxy, forming a new ring of dust, gas, and stars, which orbit around the inner galaxy almost at right angles to the larger galaxy's disk. This is the vertical polar ring which we see almost edge-on in Hubble's Wide Field Planetary Camera 2 image of NGC 4560A, created using 3 different color filters (which transmit blue, green, and near-infrared light).

The ring appears to be highly distorted and the presence of bluish, young stars below the main ring on one side and above on the other shows that the ring is warped and does not lie in one plane. The typical ages of the stars in the polar ring may provide a clue to the evolution of this unusual galaxy. Because the polar ring extends far into the halo of NGC 4650A, it also provides a unique opportunity to map "dark matter," which is thought to surround most disk galaxies.

The HST exposures were acquired by the Hubble Heritage Team and guest astronomers Jay Gallagher (University of Wisconsin-Madison), Lynn Matthews (National Radio Astronomy Observatory-Charlottesville), and Linda Sparke (University of Wisconsin-Madison). Image Credit and comments: The Hubble Heritage Team (AURA/STScI/NASA).

activesun_trace_big.jpg

This is a false-color image of an active region near the edge of the Sun, as imaged by the TRACE satellite. Hot plasma is shown exploding from the Sun's photosphere and travelling along loops defined by the Sun's magnetic field. Incidentally, these loops are so large that the entire planet Earth could easily fit beneath one of them. This image is credited to A. Title (Stanford Lockheed Institute), TRACE, and NASA.

aurora1_sts39_big.jpg

The Southern Lights (Aurora Australis) as seen from Space Shuttle Discovery flight STS-39 in April of 1991. The glow surrounding the shuttle's engines is caused by ionized atomic gas. The aurora depicted are at a height of about 50-80 miles. This image is credited to the STS-39 Crew and NASA.

boulder_a17_big.jpg

A boulder on the Moon as depicted by Eugene Cernan during the Apollo 17 mission, the last manned Moon landing. Standing next to the split boulder is scientist-astronaut Harrison Schmitt. Apollo 17 explored the terrain at the Taurus-Littrow landing site, used explosives to test the internal geology of the Moon, and returned the most lunar samples of any mission. Credit for this image goes to Eugene Cernan, the Apollo 17 Crew, and NASA.

caloris_m10.gif

The Caloris Basin is the largest surface feature on the planet Mercury. This large crater, more than 1,000 kilometers across, was created by a collision with an asteroid. The Caloris Basin is one of the hottest spots on Mercury's surface, located directly under the Sun when Mercury is closest to the sun. This image is credited to NASA, JPL, and Mariner 10.

catseye_hst_big.jpg

A Hubble Space Telescope image of the "Cat's Eye Nebula," a planetary nebula located some 3,000 light years away. Planetary nebulas are so called because they appear small, round, and planet-like in small telescopes. High-resolution images such as this, however, reveal them to be stars surrounded by shells of gas blown off during the late stages of stellar evolution. This image is credited to J.P. Harrington and K.J. Borkowski (University of Maryland), HST, and NASA.

cena_wfpc2_big.jpg

The center of the active galaxy Centaurus A is shown in this mosaic of Hubble Space Telscope images taken in blue, green, and red light. The images have been processed to present this natural color picture.

At the center of this remarkably active object, infrared images from the Hubble have shown what appear to be disks of matter spiralling into a black hole that has a billion times the mass of the Sun. Centaurus A appears to be the result of a collision between two galaxies; it is located about 10 million light-years away. This image is credited to E.J. Schreier (STScI) et al. and NASA.

centaurus.jpg

A view including the area of the constellations Centaurus and Circinus. The image was obtained with a usual mechanic camera and lens of normal focal length (50mm). Guiding was done on a Super Polaris mount. The place is the Cerro Tololo Observatory in Chile. The conventional photography was digitized to 2048 x 3072 pixels and a color depth of 3 x 8 Bits. Digital image processing improved the quality of the image dramatically. The main steps are the subtraction of the terrestrial skybackground and following contrast enhancement.

The exposure for the original image was 60 minutes on Scotchchrome 400 film with a 50mm f1.4 lens.

In mid-May Centaurus culminates at about 22:00 (10 pm) local time. The galactic equator is crossing our field of view with bright star clouds and in between dark fields of interstellar dust that obscure the background starlight, most prominent the so called Coal Sack (Crux, right). A lot of star clusters join this field (see the deep sky objects). At the upper right is the Globular Star Cluster Omega Centauri. It is the brightest Globular Cluster in the sky, whereas it appears too bright on the photography because of its large angular dimension of more than half a degree. To the naked eye, it is a diffuse star of 3.6th magnitude. Photograph © Till Credner and Sven Kohle.

chamaeleon_vlt_big.jpg

A photogenic group of nebulae can be found in the southern constellation Chamaeleon. Towards Chamaeleon, dark molecular clouds and bright planetary nebula NGC 3195 can be found. Visible near the center of the photograph is a reflection nebula surrounding a young bright star. On the lower right, a dark molecular cloud blocks

the light from stars behind it. Image and comments credit: FORS
Team, 8.2-meter VLT Antu, ESO.

columbia_jan96_big.jpg
The launch of the Space Shuttle Columbia in January 1996 from pad
39-A at the Kennedy Space Center in Florida. Photo credit: NASA.

copernicus_apollo17_big.gif
The crater Copernicus is one of the most prominent on the Moon,
easily visible with binoculars slightly northwest of the center of the
disk. About 93 kilometers across, Copernicus is seen here from space
in this 1972 photograph taken during the Apollo 17 mission. Credit
for this image goes to the Apollo 17 Crew and NASA.

crab_hst_big.jpg
At the heart of the Crab nebula (M1), the result of a supernova
explosion in the summer of 1054AD, is the core of the collapsed
star—a rotating neutron star or pulsar. This object, only 6 miles wide
but more massive than our Sun is the left most of the two bright
central stars in this Hubble Space Telescope image. This photograph
is credited to J. Hester and P. Scowen (ASU), and NASA.

cygnusloop_hst_big.gif
The shockwave from a star that exploded 15,000 years ago in the
constellation Cygnus is still expanding into space. As the gas collides
with a stationary cloud, it is heated and begins to glow in both visible
and high energy radiation. This glowing cloud is known as the
Cygnus Loop (NGC 6960/95), and is located about 2,500 light-years
away. The colors in this Hubble Space Telescope image represent
emissions from different kinds of atoms: oxygen is blue, sulfer is red,
and hydrogen is green. Credit for the image goes to J. Hester (ASU)
and NASA.

darkcloud_2massatlas_big.jpg
Feitzinger and Stuwe 1-457 is a nearby fuliginous nebula some 1,000
light-years away. It is seen here against a star field in our home galaxy.

Credit for this image goes to the Two Micron All Sky Survey Project (IPAC/U. Mass.).

deepfield_ntt_big.jpg
A deep field image from an Earth-based telescope, the New Technology Telescope, shows galaxies as faint as 26th magnitude. This image was the result of pointing the telescope at a small section of the sky that contained no bright objects at all. This image is credited to S. D'Odorico et al., NTT, and ESO.

dem192_umich_big.jpg
A small section of a star forming region (DEM192) in the Large Magellanic Cloud, this picture is a composite of three separate photographs, each sensitive to only one color of light that distinguishes a specific chemical element. This image is credited to C. Smith (University of Michigan), Curtis Schmidt Telescope, CTIO, Chile.

earth.gif
Perhaps one of the most evocative images from the era of manned lunar exploration, this image of the whole Earth suspended against the darkness of space was photographed from Apollo 17. It is NASA image AS17-148-22727.

eclipse77b.gif
An image of a total solar eclipse in 1977(or thereabouts), this image is courtesy of SEDS.LPL.Arizona.EDU.

eclipse-setting.gif
An unusual photograph of an annular solar eclipse in 1992, shows the eclipsed Sun setting on the horizon. Annular eclipses are caused when the Moon's distance from the Earth is great enough that its disk only covers part of the disk of the Sun. When the Moon and Sun are aligned, the result is a "ring of fire" such as the one shown in this image. Image courtesy of SEDS.LPL.Arizona.EDU.

europa3_vg_big.jpg

The dark streaks visible in this Voyager mission image of the surface of Jupiter's moon Europa may be cracks in the ice covered surface of an ocean. The surface characteristics of Europa are suggestive of sea ice on Earth. Such cracks may be caused by Jupiter's tidal stresses, accompanied by freezing and expansion of an underlying layer of water. This image is credited to the Voyager Project, JPL, and NASA.

europacolor_gal_big.jpg

This color image of the surface of Jupiter's moon Europa, shows an area of about 120 by 150 miles. In this image, blue tints represent relatively old ice surfaces, while reddish regions may contain material from more recent geological activity. The white splotches in the image are bright materials blasted from crater Pwyll, about 600 miles south (to the right). This image is credited to the Galileo Project, the University of Arizona, JPL, and NASA.

europaridges_gal_big.jpg

There is speculation that the wealth of ridges might be caused by cold water volcanoes, and that beneath the ice there is a liquid ocean. This image shows common features on Europa's surface—pure blue water ice beneath lighter ridges. These ridges might be caused by volcanic activity. The reasons for the colors of the ridges remains unknown. Photo credit goes to The Galileo Project, JPL, and NASA.

farside_apollo16_big.gif

A picture of the far side of the Moon from the Apollo 16 mission. Since the Moon is locked in synchronous rotation, one side always faces the Earth, the far side can only be seen from space. Much of the side of the Moon we normally see is covered with smooth, dark lunar maria, but the far side is much more rough. One explanation for the difference between the two sides is that the far side crust is thicker, thus making it more difficult for lava to flow from the interior and form the smooth maria. Image credited to Apollo 16 and NASA.

feix0424_trace.jpg

A colorized image looking across the edge of the Sun was made by the TRACE satellite in the extreme ultraviolet light radiated by highly ionized iron atoms. It shows arcs of intensely hot gas suspended in powerful looping magnetic fields. The Sun's surface is a relatively cool 6,000 degrees Celsius, but the hot plasma loops reach about a million degrees. This image is courtesy of CFA, the TRACE Team, and NASA.

gaspra_gal_big.gif

Asteroid 951, also known as Gaspra, is a chunk of rock about 20 kilometers long that orbits the Sun in the main asteroid belt between Mars and Jupiter. The colors of this image have been exaggerated to highlight changes in reflectivity, surface structure, and composition. This image is credited to The Galileo Project, NASA.

gc2.jpg

Appearing as if it might be one of the great canyon structures on the planet Mars, this image of the Grand Canyon in Arizona was photographed from Space Shuttle mission STS61 in 1985. Image credit goes to the STS-61 Crew, NASA.

goes2.gif

An unusual view of a total solar eclipse, the dark spot visible on the Earth is the Moon's shadow as seen by the GOES Earth satellite during the November 7, 1991 eclipse. This image is courtesy of SEDS.LPL.Arizona.EDU.

gularif1_magellan_big.gif

A computer-generated image of a rift valley on Venus, created from data from the Soviet Venera landers and radar data from the Magellan spacecraft. In the foreground is the edge of a rift valley created by crust faulting; the valley runs to the base of Gula Mons, a two mile high volcano seen on the right—some 450 miles in the distance. Another Venusian volcano, Sif Mons, appears on the left. The Magellan spacecraft used radar to image the surface of Venus through the constant cloud cover. The spacecraft was able to image over 98 percent of the Venusian surface. This image is credited to The Magellan Project, JPL, and NASA.

halebopp5_aac_big.jpg

Comet Hale-Bopp, one of two bright comets to recently grace Earth's skies, appears over Val Parola Pass in the Dolomite mountains surrounding Cortina d'Ampezzo, Italy. This photograph shows Comet Hale-Bopp's two tails: The blue ion tail is created by fast moving particles from the solar wind directly expelling ions from the comet's nucleus, and the whitish dust tail is composed of particles of dust and ice expelled by the comet's nucleus. This beautiful image is credited to A. Dimai (Col Druscie Observatory), ACC.

halebopp6_aac_big.jpg

Comet Hale-Bopp, showing another view of the spectacular ion and dust tails. Although cometary tails are spectacular to view, they are so tenuous that they are actually very nearly perfect vacuums. This photo is credited to A. Dimai and D. Ghirardo (Col Druscie Observatory), ACC.

halley_hmc_big.gif

The nucleus of Halley's Comet, our first view of what a comet nucleus really looks like. This image is a composite of hundreds of photographs made by the Giotto spacecraft during its 1986 to the famous comet. Photo credit goes to the Halley Multicolour Camera Team, Giotto, and ESA. This image is copyright by MPAE.

hao_wlcc_941103.jpg

This image of the total solar eclipse of November 3, 1994 was photographed from Putre, Chile by a research team from the High Altitude Observatory of the National Center for Atmospheric Research in Boulder, Colorado. The project was sponsored by the National Science Foundation. The photograph was taken with a special camera system developed by Gordon A. Newkirk, Jr., that permits more delicate detail to be recorded than conventional photographic methods would show.

hawaii_sts26_big.jpg

The Hawaiian Island archipelago as seen from the Space Shuttle Discovery in October 1988. The peak of Mauna Kea on the Big Island (upper left) sports an impressive assortment of major telescopes, including the twin Keck instruments. Photo credits to the STS-26 Crew, NASA.

helixd_hst_big.gif

Droplet-like condensations nicknamed "cometary knots" are easily visible in this detail of Helix nebula photographed by the Hubble Space Telescope. The individual knots are several billion miles across, roughly twice the size of our entire Solar System, while the tails stretch out over 100 billion miles. The Hubble Space Telescope has shown indications of literally thousands of these structures. It is speculated that hot, fast moving shells of nebular gas overrunning cooler, denser, slower shells ejected by the central star during an earlier expansion might cause these condensations as the two shells intermix. It is possible that instead of dissapating over time, these knots could collapse and form small icy worlds. This image is credited to R. O'Dell and K. Handron (Rice University), and NASA.

ida_dactyl.gif

This color photograph is created from images made by the Galileo spacecraft about 14 minutes before its closest approach to asteroid 243 Ida on August 28, 1993. When imaged, Ida was about 6,500 miles from the spacecraft. This is a false-color image—a "true color" image would appear mostly gray. This photographs contains what was a surprise to scientists—Ida's moon Dactyl, visible to the right of the asteroid. Dactyl is not identical in spectral properties to any area of Ida visible in this view, although the overall similarity suggests that it is made of basically the same rock types. This image is credited to the Galileo Project, JPL, and NASA.

io970321_gal_big.jpg

Io, one of the moons of Jupiter discovered long ago by Galileo, is the most volcanically active body in our solar system. This image is generated from data gathered in 1996 by NASA's Galileo spacecraft, and is centered upon the side of Io that always faces away from Jupiter. The color and contrast of the image have been enjanced to emphazide Io's surface brightness and color variations; features as small as 1.5 miles across are visible. Unlike many of the other moons of Jupiter and the other "gas giants" of the outer Solar System, no impact craters are visible on Io's surface. This indicates that the surface is regularly covered with new volcanic deposits more rapidly than craters are created. It is likely that the volcanic activity of Io is a result of the changing gravitational tides caused by Jupiter and its other large moons, which would heat the interior of Io. This image is credited to The Galileo Project, JPL, and NASA.

iohires_gal_big.jpg
This high-resolution image is of Jupiter's moon Io, the most volcanically active body in our Solar System. This image shows detail as small as three kilometers. As might be expected from the volcanic nature of Io, sulfur compounds cause many of its striking colors, while its darker regions are probably composed of silicate rock. This image is credited to the Galileo Project, JPL, and NASA.

iojup5mil_voyager_big.jpg
Io, seen against the clouds of Jupiter from some 5 million miles distant. This black and white image is from the NASA Voyager 1 spacecraft was taken during its approach to Jupiter in 1979. Credit: Voyager Project, JPL, NASA.

ioshadowc_hst_big.jpg
Orbiting Jupiter once every 43 hours, the volcanic moon Io cruises 500,000 kilometers above swirling, banded cloud-tops. Orbiting Earth once every 1.5 hours, the Hubble Space Telescope watched as Io accompanied by its shadow crossed the face of the reigning gas giant planet in 1997. This and other sharp false-color images have recently been chosen to celebrate the ninth anniversary of the Hubble's launch (April 24, 1990). Reflective patches of sulfur dioxide

"frost" are visible on Io's surface while Io's round dark shadow is seen passing over brownish white regions of Jupiter's high altitude haze and clouds. In October and November of 1999, the Galileo spacecraft currently operating in the Jovian system is scheduled to make two daring close approaches to Io, possibly flying through a volcanic plume. Credit: John Spencer (Lowell Observatory), NASA.

jupclouds_vgr_big.jpg

The clouds of Jupiter, as shown by this image from the Voyager I flyby of 1979. It is not known what causes the coloration of Jupiter's cloud surface, but one theory is that colorful hydrogen compounds might well up from the warmer regions of the Jovian atmosphere and tint the cloud tops. Another theory speculates that compounds of trace elements such as sulfur may color the clouds. Jupiter's colors *do* indicate the clouds' altitudes, with blue being the lowest and red the highest. The intricate vortices shown here were first revealed in the flyby of Voyager I. Credit for this image goes to the Voyager Project, JPL, and NASA.

jupring1_gal_big.jpg

Among the many discoveries made by the Voyager spacecrafts as they made their way through the outer solar system were the rings of Jupiter. Voyager I during its 1979 flyby discovered these rings, but their origin was unknown. Recent observations by the Galileo spacecraft confirm that these rings are created by meteoroid impacts on small nearby moons. This image from the Galileo spacecraft is credited to M. Belton (NOAO), J. Burns (Cornell) et al., Galileo Project, JPL, and NASA.

keyhole_noao.jpg

NGC3324, the Keyhole nebula, is named for its unusual shape. The Keyhole nebula is a small region superposed on the larger Eta Carina nebula, both of which were created by violent outbursts of the dying star Eta Carina. The Keyhole nebula is an emission nebula that contains much dust. It is approximately 9,000 light years distant. Southern Hemisphere observers can easily see this nebula, even with a small telescope.

The Keyhole Nebula was recently discovered to contain highly structured clouds of molecular gas. This image is credited to NOAO, and is copyrighted by AURA, with all rights reserved.

lagoonclose_hst_big.jpg

This spectacular Hubble Space Telescope image shows a detail of dusty molecular clouds in the Lagoon nebula (M8), located about 5,000 light-years distant in the constellation Sagittarius. In this star-forming region at least two long funnel-shaped clouds can be seen, each roughly a half light-year long. It is possible that these huge structures are directly analagous to Earth's tornadoes, formed in this case by extreme stellar winds and intensely energetic starlight pouring into regions of cool gas and dust. It's definitely not Kansas. This image is credited to A. Caulet (ST-ECF, ESA) and NASA.

lg_earthrise_apollo8.gif

Perhaps the most famous photograph from the age of manned space exploration, this image of the Earth rising over the desolate lunar limb was recorded during the flight of Apollo 8. Frank Borman, James Lovell, and William Anders were the crew of Apollo 8, which launched on December 21, 1968, completed ten lunar orbits, and landed December 27, 1968. This famous image is credited to NASA and the Apollo 8 Crew.

loop1_rosat_big.gif

One of the largest coherent structures on the sky is known simply as Loop I and can best be seen in radio and X-ray maps. Spanning over 100 degrees, part of Loop I appears so prominent in northern sky maps that it is known as the North Polar Spur (NPS). Loop I, shown above in X-ray light, is a thin bubble of gas about 350 light-years across with a center located only about 400 light-years away. Surprisingly, the cause of this immense structure is still debated, but is possibly related to expanding gas from a million-year old supernova. Loop I gas is impacting the nearby Aquila Rift molecular cloud, and may create relatively dense fragments of the local interstellar medium. Were our Sun to pass through one of these fragments in the next few million years, it might affect Earth's climate. Credit and comments: ROSAT, MPI, ESA, and NASA.

lunarscape_apollo17_big.gif

A photograph of astronaut Harrison Schmitt next to the lunar roving vehicle, taken by Eugene Cernan during the Apollo 17 mission. Family Mountain is in the center background, and the edge of South Massif is at the left of the frame. Image credited to Apollo 17 and NASA.

lunmod_a15_big.gif

A home away from home, the lunar module "Falcon" sits on the Moon's surface, with the Apennine mountains in the background and Mount Hadley Delta visible to the right. This image was taken during the Apollo 15 mission in July/August 1971. This image is credited to the Apollo 15 Crew and NASA.

lunrov_a15_big.gif

Getting ready for a drive on the Moon, astronaut James Irwin works on the first Lunar Roving Vehicle during the Apollo 15 mission. Behind the lunar module "Falcon" on the left are lunar mountains Hadley Delta and Apennine Front. Photograph credits go to Apollo 15, David Scott, and NASA.

m16_hst_big.jpg

Considered by many to be among the most striking images yet produced by the Hubble Space Telescope, these image shows star forming regions called EGGs (Evaporating Gaseous Globules) at the ends of giant pillars of gas and dust in M16, the Eagle nebula. This picture was taken by Hubble's Wide Field and Planetary Camera, and is credited to J. Hester and P. Scowen (ASU), HST, and NASA.

m100Color.gif

A photograph of the galaxy M100 in the constellation Coma Berenices, taken by the Hubble Space Telescope. M100 is one of 2,500 or so members of the Virgo galaxy cluster—amateur astronomers can see this face-on galaxy as a faint object. The Hubble

Space Telescope can image this distant galaxy in such detail, however, that it is possible to observe Cephied Variables and thus accurately measure its distance from us. This image is credited to J. Trauger, JPL, and NASA.

m27_vlt_big.jpg

Popularly known as the "Dumbbell nebula," M27 is a planetary nebula some 1,200 light-years away from us in the constellation Vulpecula. This false color image was produced during testing of the European Southern Observatory's Very Large Telescope. This image is credited to the FORS Team, 8.2-meter VLT, and ESO.

m31gc1_hst_big.jpg

While we may be familiar with some of the striking photographs of our "local" globular clusters, this image shows something of the power of the Hubble Space Telescope as a research instrument. This globular cluster, known as both G1 and Mayall II, is a companion to the famous M31 galaxy in Andromeda, and contains over 300,000 stars. Credit for this remarkable image goes to M. Rich, K. Mighell, and J.D. Neill (Columbia University), W. Freedman (Carnegie Observatories), and NASA.

m45.gif

The famous Pleiades, also known as the Seven Sisters and M45, contains over 3,000 stars. This naked-eye cluster in the constellation Taurus is about 400 light-years away, and 13 light-years across. One of the brightest and nearest open star clusters, M45 is shrouded in a blue reflection nebula that can be seen around the brighter stars in this image. Faint brown dwarf stars have recently been found in the Pleiades. This photograph is credited to David Malin (AA), AATB, ROE, and the UKS Telescope.

m6_noao_big.jpg

M6, also known as the butterfly cluster and NGC 6405, is an open cluster of stars about 20 light-years wide and 2,000 light-years distant. Located in the constellation Scorpius, M6 is an easy object for binoculars. This open cluster, like many other such open clusters, is primarily composed of young blue stars, although the brightest star in the cluster is distinctly orange. This image is credited to AURA, NOAO, and NSF.

m9609c.jpg

A photograph of the September 26-27, 1996 lunar eclipse, as photographed by Sky & Telescope's Rick Fienberg. Rick photographed this eclipse from suburban Boston, Massachusetts using his Meade 8-inch f/10 Schmidt-Cassegrain telescope and ISO 200 color-print film.

mars_surface_vik2_big.jpg

A ground-level view of Utopia Planitia (the Plain of Utopia) on the planet Mars, as imaged by Viking 2 in August 1976. Near the center of this image are shallow trenches made by Viking 2's soil sampler arm. Photo credit: The Viking Project, M. Dale-Bannister (WU StL), and NASA.

marsglobe1_viking_big.jpg

An image of the planet Mars from space, created by the Viking spacecraft. Credits for this image go to the USGS, The Viking Project, and NASA.

marshill_mpf_big.jpg

The robot rover Sojourner analyzes a rock named Moe on the surface of Mars. In the foreground of this image is a part of the Sagan Memorial Station, named for the late Carl Sagan, and the hill about 1 kilometer in the distance is "Twin Peaks." This image is credited to the IMP Team, JPL, and NASA.

marsnpole_mola_big.jpg

A 3D visualization of Mars' north pole, this image is based on elevation measurements from an orbiting laser. This polar ice-cap is about 1,000 km across, a maximum of 3 km deep, with canyons and troughs up to 1 km deep. Mars' polar cap is composed primarily of water ice, but contains only about four percent of the Earth's Antarctic ice sheet. Image credit goes to Greg Shirah (SVS), MOLA Team, MGS Project, and NASA.

marsval_mgs_big.jpg
marsval1_mgs.jpg

A tiny part of the northern edge of the canyon Valles Marineris on the planet Mars, the large image was created from high resolution Mars Global Surveyor images that were combined with color data from the Viking Orbiter. Valles Marineris stretches some 2,500 miles across the surface of Mars; the area pictured in the larger image is about six miles wide, and details 20 to 30 feet across can be seen. A detail of the large image is included here as a separate file, marsvall_mgs.jpg. These striking images are credited to the MOC Team, MGS, and NASA.

massclus_hst_big.jpg

Each of the fuzzy spots in this false color Hubble Space Telescope image of a newly discovered galaxy cluster is a galaxy similar in mass to our home galaxy. This particular cluster of galaxies is a few million light-years across, and some 8 *billion* light-years distant. Image credit: M. Donahue (STScI) et al., NASA.

merctran.gif

A view from space of a rare event, the transit of the planet Mercury across the face of the Sun. This photograph was taken in soft x-ray light (3-45 Angstroms) by Yohkoh, a Japanese satellite designed to study solar flares and coronal disturbances on the Sun. For reference purposes, the transit track is drawn on the left-hand image. Photo credit: ISAS (Institute of Space and Aeronautical Sciences).

mgn_c115n266_1.gif

Wheatley Crater on the planet Venus is shown in this synthetic aperture radar image from the Magellan spacecraft. Located in the Asteria Regio, Wheatley Crater is 72 km in diamater. The synthetic aperture radar on Magellan was able to penetrate the constant dense cloud cover of Venus to map about 98 percent of the planet's surface in great detail. This image is a portion of Magellan C1-MIDR 15N266, framelets 21 and 22. Image Credit: the Magellan Team, NASA.

mgn_c130s279_1.gif

A Magellan spacecraft image of a nova, a radial network of grabens, in Themis Regio, on the surface of the planet Venus. About 50 novae have been identified on Venus, each consisting of closely spaced grabens radiating from a central area. This nova is about 250 km in diameter, concentrated to the south. (North is up.) This image is Magellan C1-MIDR 30S279;1, framelet 18. Image Credit: the Magellan Team, NASA.

moc2_107_msss.gif

Ripples on the cratered terrain north of Hesperia Planum, from a Mars Orbiter Camera view. This ancient, cratered surface sports a covering of windblown dunes and ripples oriented in somewhat different directions. The dunes are bigger and their crests generally run east-west across the image. The ripples are smaller and their crests run in a more north-south direction. This picture covers an area only 3 km (1.9 miles) wide. Illumination is from the top.

moc2_108_msss.gif

The Olympica Fossae are a collection of troughs and depressions located in northern Tharsis, south of the Alba Patera volcano. The Mars Global Surveyor Mars Orbiter Camera has been sending back unprecedented, spectacular views of this region. The Olympica Fossae are especially interesting because they show landforms that run the entire range of things seen elsewhere on Mars. This picture shows many examples, including layered outcrops in canyon walls, evenly-spaced dunes on the canyon floors, dark landslide streaks on the canyon walls, pits formed by ground collapse, and streamlined forms related to the flow of water, mud, or lava.

moc2_119_msss.jpg

A patch of bright clouds hanging over the summit of the Apollinaris Patera volcano, on Mars, as photographed by the Mars Global Surveyor Mars Orbiter Camera in April 1999. The caldera at the volcano's summit is about 50 miles (80 km) across.

The color in this picture was derived from the MOC red and blue wide angle camera systems and does not represent true color as it would appear to the human eye (that is, if a human were in a position to be orbiting around the red planet).

moonmir_sts91_big.jpg

Photographed in June 1998 from the Space Shuttle Discovery during mission STS-91, the Mir Space Station and the Moon appear above thick storm clouds on the Earth's surface. This photograph is credited to the STS-91 Crew and NASA.

mosaicfull.jpg

The "deepest ever" view of the universe, this Hubble Deep Field image covers a tiny portion of the sky about 1/30 the diameter of the full moon. The field is so narrow that only a few foreground in our local galaxy are visible—the relatively bright object with diffraction spikes just left of the center of the image may be a 20th magnitude star. Some of the galaxies in this image are as faint as 30th magnitude—some 4 billion times fainter than can be seen with the human eye.

This true color image is a mosaic of 276 frames taken by the Hubble Space Telescope between December 18 and the 28, 1995. A total of 342 frames were taken, but only 276 had been fully processed by the date this image was released. Credit for this image goes to Robert Williams, the Hubble Deep Field Team (STScI) and NASA.

n81_hst_big.jpg

An image of N81, a young star cluster in the Small Magellanic Cloud, some 200,000 light-years distant. This region is an active star forming region. Image credit: M. Heydari-Malayeri (Paris Observatory) et al., WFPC2, HST, and NASA.

neptunex.gif

The Great Dark Spot on Neptune, as photographed by NASA's Voyager 2 spacecraft in August 1989. When Neptune was viewed by the Hubble Space Telescope in 1994, the spot had vanished, replaced by another dark spot in the planet's northern hemisphere. Winds near the Great Dark Spot were measured up to 1,500 miles per hour, the strongest winds recorded up any planet. Credit for this image goes to the Voyager Project, JPL, and NASA.

ngc2440_hst2_big.jpg

The planetary nebula NGC 2440, contains one of the hottest white dwarf stars known, seen here as the bright dot near the center of the photograph. Early in its evolution, a white dwarf star casts off a shell of gas, as shown in this Hubble Space Telescope image. This photograph is credited to H. Bond (STScI), R. Ciardullo (PSU), WFPC2, HST, and NASA. This false color image was post-processed by F. Hamilton.

ngc4945_mpg_big.jpg

NGC 4945 is a spiral galaxy in the Centaurus Group of galaxies, located only six times further away than the prominent Andromeda Galaxy. The thin disk galaxy is oriented nearly edge-on, however, and shrouded in dark dust. Therefore galaxy-gazers searching the southern constellation of Centaurus need a telescope to see it. This picture was taken with a large telescope testing a new wide-angle, high-resolution CCD camera. Most of the spots scattered about the frame are foreground stars in our own Galaxy, but some spots are globular clusters orbiting the distant galaxy. NGC 4945 is thought to be quite similar to our own Milky Way galaxy. X-ray observations reveal, however, that NGC 4945 has an unusual, energetic, Seyfert 2 nucleus that might house a large black hole. Credit: 2P2 Team, WFI, MPG/ESO 2.2-m Telescope, La Silla, ESO.

ngc604_hst_big.jpg

The nebula NGC 604 is a star forming region some 1,500 light-years across in the galaxy M33. In this photograph by the Hubble Space Telescope, a gaseous interstaellar cloud surrounds more than 200 newly formed stars. This photograph is credited to H. Yang (UIUC), J. Hester (ASU), and NASA.

ngc6210_hst_big.jpg

The planetary nebula NGC 6210 in the constellation Hercules is about 6,500 light-years distant. This true color image from the Hubble Space Telescope shows hot jets of gas streaming through holes in an older and cooler shell of gas, as well as the central star that created the nebula (inset image). Photo credit: R. Rubin (NASA/Ames) et al., WFPC2, HST, and NASA.

ngc7635big.jpg

Part of NGC 7635, the Bubble nebula, as imaged by the Hubble Space Telescope. The entire nebula is about ten light-years across. The glowing gas in the lower right-hand corner of this image is a dense region of material that is being shaped by strong stellar winds and radiation from the massive bright star at the left of the frame. Photo credit: Hubble Heritage Team (AURA/STScI/NASA).

ngc7742big.jpg

A face-on view of the small spiral Seyfert 2 active galaxy NGC 7742, as photographed by the Hubble Space Telescope. Seyfert 2 active galaxies are probably powered by a black hole at their center. Photo credit: Hubble Heritage Team (AURA/STScI/NASA).

northmoon_gal_big.jpg

Impact craters on the Moon, as photographed by the Lunar Prospector spacecraft. Image Credit: Galileo Project, JPL, NASA.

oberon2_usgs_big.gif

Oberon, second largest and most distant of the moons of Uranus, processed from images by the Voyager 2 spacecraft during its 1986 flyby. This image was constructed by the U.S. Geological Survey (USGS), and represents all known surface features. The large dark crater to the right of center in this image is named Hamlet. Image credit: Astrogeology Team (USGS), the Voyager Project.

ocsat_anim.gif

On September 18, 1997, many stargazers in the U.S. were able to watch a lovely early morning lunar occultation as a bright Moon passed in front of Saturn. Using a 1.2 meter reflector, astronomer Kris Stanek had an excellent view of this dream-like event from the Whipple Observatory atop Arizona's Mount Hopkins. This animated .gif image was constructed by Wes Colley from 4 frames taken by Stanek at 35-second intervals as the ringed planet emerged from behind the Moon's dark limb. While lunar occultations of fairly bright stars and planets are not extremely rare events, their exact timing depends critically on the observer's location. Credit and copyright: K. Stanek (CfA), W. Colley (Princeton).

omegacen_umi_big.gif
The globular cluster Omega Centauri is the largest of about 160 such clusters known in our galaxy, containing about 10 million stars. Photo credit: P. Seitzer (University of Michigan).

ori2mass_big.jpg
A mosaic of the stellar nursery in the Orion nebula, constructed using data from the 2 Micron All Sky Survey (2MASS). 2MASS cameras are sensitive to near infrared light around 2 microns, thus the name. Survey observations in three infrared bands were converted to blue, green, and red color information to produce this image. Image credit goes to the 2MASS Collaboration, University of Massachusetts, and IPAC.

orion-big.jpg
The constellation Orion is rich in objects of interest to astronomers, amateur and professional alike. In this 60-minute exposure on Kodak Ektachrome 400 Elite through a 55mm f/3.5 lens, the great Orion Nebula (M42/43) can be seen in Orion's "sword," hanging below the three stars making up the belt. The nebulosity immediately to the left of the left-most belt star in this image is the Flame nebula, and the reddish glow located immediately below this star is where the Horsehead nebula (IC 434) is located. The brightest star in Orion is Rigel, visible at the lower right of this image, the second brightest is the prominent reddish-orange Betelgeuse, which makes up Orion's

left shoulder. Photograph taken by Till Credner and Sven Kohle, © "Observers," Astronomical Institutes of the University of Bonn.

orion_iras_big.jpg

An unfamiliar view of the constellation Orion, this picture was created using data from IRAS, the InfraRed Astronomical Satellite. Using information recorded at three invisible infrared wavelengths and then assigning the colors red, green, and blue to them created this false color image covering a field about 30 x 24 degrees. The star Betelguese appears as a small purplish dot just above the center of this image, and the famous M42/43 complex appears as the yellow areas at the lower right. Photo credit goes to the IRAS Project, IPAC.

orionuwh3_avann_big.jpg

A large section of a winter night sky, this photograph from the island of La Palma, in the Canaries archipelago. The constellation Orion is easy to pick out, the star Sirius is near the bottom of the image, and the v-shaped cluster of yellowish stars in the upper right part of this image is part of the constellation Taurus. Photograph credit and copyright: A. Vannini, G. Li Causi, A. Ricciardi, and A. Garatti.

pan_enhanced.gif

An enhanced, geometrically improved, and color enhanced version of the 360-degree panorama from the Imager for the Mars Pathfinder (IMP). The IMP is a stero imaging system that, when fully deployed, stands 1.8 meters above the Martian surface and has a resolution of two millimeters at a range of 2 meters. The color of this image is enhanced both in saturation and intensity.

"Twin Peaks" is visible on the horizon, about 1-2 kilometers, and the rock "Couch" is the dark curved rock at right of "Twin Peaks." The tracks of the Sojourner are clearly visible in this high-resolution image. Photo credit for this image goes to the Mars Pathfinder Team, JPL, and NASA.

pegdsph_grebel_big.gif

The Pegasus dwarf spheriodal is a recently recognized member of our local group of galaxies. The fainter, bluer stars that make up this small galaxy are almost hidden by glare of brighter foreground stars in our own galaxy. The "spikes" on the brighter stars in this photograph are imaging artifacts; this image was made with the Keck telescope. Image credit goes to E. Grebel (U. Washington) and P. Guhathakurta (UCO/Lick).

phobos.gif

The larger of the two moons of Mars, Phobos is about 17 miles across and orbits Mars in less than 8 hours. Its orbit is about 3,600 miles from the Martian surface, close enough that gravitational tidal forces are dragging it down. Eventually Phobos will either crash into the surface of Mars or be shattered by tidal forces—forming a ring around the red planet. This photograph from the Viking Orbiter, was produced in 1977. Photo credited to NASA/JPL and the Viking Project.

phot-15-99-preview.jpg

PR Photo 15/99 is a color a composite of three individual exposures, taken with FORS1 at the VLT UT1 on February 3, 1999. They were obtained through B (blue), V (green), and H-alpha (red) filters, and each of them was bias-subtracted and flat-fielded before combination. The field-of-view is 6.8 x 6.8 arcmin2, corresponding to an area of just over 300 x 300 light-years2 at the distance of the Large Magellanic Cloud. The exposure times were 60, 60, and 450 seconds and the image quality was 1.0, 0.7, and 0.8 arcsec (FWHM) for the B, V, and H-alpha frames, respectively. The intensity scale is linear for the B and V images and logarithmic for H-alpha, in order to enhance the fine structure of the faintest parts of the nebulosity. North is 33° to the right of the vertical.

PR Photo 15/99 is a composite of blue and green images plus an image in the light of Hydrogen. It shows that there is indeed still much gas around NGC 1850.

While part of this may well be the remnant of the "parent" gas cloud (i.e. the one from which both clusters were born), the presence of filaments and of various sharp "shocks," e.g. to the left and below NGC 1850, offers support to the theory of supernova-induced star birth in the younger of the two clusters. Some "protostars" are located near or in some of the filaments—this is interpreted as additional evidence for that theory.

The nebulosity directly above the main cluster that is shaped like a "3" is the well-known supernova remnant N57D which itself may also be associated with NGC 1850. Image and comments courtesy of the FORS Team, 8.2-meter VLT, and ESO.

pillars3_hst_big.jpg

Evaporating gaseous globules (EGGs) are star forming areas in M16, the Eagle nebula, about 7,000 light-years distant. This striking image from the Hubble Space Telescope shows the EGGs at the tips of giant, light-years long pillars of dense gas. Photo credit: H. Hester, P. Scowen (ASU), HST, and NASA.

rabaul.gif

An eruption of the Rabaul Caldera in Papau New Guinea, as photographed from the Space Shuttle in September 1994. Photo credit goes to the STS 64 Crew and NASA.

rcw108_eso_big.jpg

Hot blue stars, red glowing hydrogen gas, and dark, obscuring dust clouds are strewn through this dramatic region of the Milky Way in the southern constellation of Ara (the Altar). About 4,000 light-years from Earth, the stars at the left are young, massive, and energetic. Their intense ultraviolet radiation is eating away at the nearby star forming cloud complex—ionizing the hydrogen gas and producing the characteristic red "hydrogen-alpha" glow. At right, visible within the dark dust nebula, is a small cluster of newborn stars. This beautiful color picture is a composite of images made through blue, green, and hydrogen-alpha filters. Credit: WFI, European Southern Observatory.

sataurora_hst_big.jpg

This remarkable Hubble Space Telescope image of Saturn, the second largest planet in our Solar System, shows pole-encircling Saturnian ultraviolet auroral diplays. Saturn's aurora towers more than 1,000 miles above Saturn's cloud tops. This is a false color image in which red identifies emissions from atomic hydrogen and the more concentrated white areas are due to hydrogen molecules. Photo credit: J. Trauger (JPL) and NASA.

satmoons_vg1_big.gif
The ringed planet Saturn and two of its larger moons, Tethys and Dione, appear in this photograph from the Voyager 1 spacecraft in November 1980. Photo credit: Voyager 1, JPL, and NASA.

saturn_24Apr_hst_big.jpg
Periodically the rings of Saturn appear edge-on as seen from the Earth, as in this Hubble Space Telescope image taken on August 6, 1995. Titan, Saturn's largest moon, is the brownish globe at the left of the image—its shadow can be seen on the planet's cloud tops. Four other Saturnian moons can be seen just above the the ring plane. These are, from left to right: Mimas, Tethys, Janus, and Enceladus. The dark band across the planet is actually the shadow of the famous rings. Credit for this image goes to Erich Karkoschka (University of Arizona Lunar & Planetary Lab) and NASA.

saturn_down_vg1.gif
This evocative picture of Saturn could not possibly have been taken from Earth. Taken by the Voyager 1 spacecraft in November 1980, the camera is looking back toward the Sun at the night side of Saturn, and shows the planet's shadow on the complex ring system. Since the Earth is an "inner planet" from Saturn, we are only able to see the Sun side of the ringed planet. This photograph is credited to the Voyager Project, JPL, and NASA.

saturn_vg2_big.jpg
The planet Saturn and two of its moons, Rhea and Dione, are visible (the faint dots at the right and lower right) in this image from the Voyager 2 spacecraft as it flew by the ringed planet in 1981. Photo credit: Voyager 2 and NASA.

scutum.jpg
The constellation Scutum is shown in the center of the image with Serpens Cauda right above. Below is a part of Sagittarius. The

constellation Serpens is divided into two parts: Serpens Caput and Serpens Cauda, the head and the tail of the serpent. In between is the constellation Ophiuchus, the Serpent Bearer.

In mid-August, Scutum culminates at about 21:00 LT (9pm). The galactic equator diagonally crosses the field of view in this photograph, where bright star clouds can be seen. The Scutum Star Cloud is visible to the naked eye. The number of galactic deep sky objects in this region is remarkable since it is so close to the galactic center. A part of the central region of our galaxy can be seen in the lower right, the so-called galactic bulge. Photograph taken by Till Credner and Sven Kohle, © "Observers," Astronomical Institutes of the University of Bonn, Germany.

sgr1_hst_big.jpg
A multitude of star colors appears in this Hubble Space Telescope photograph of the Sagittarius Star Cloud. Among this collection are some of the oldest stars known. This image is credited to the Hubble Heritage Team (AURA/STScI/NASA).

sl9g_hst_big.gif
A recent cataclysmic event in our solar system, the impact of a string of comet pieces on the planet Jupiter was the object of much scientific interest and scrutiny. This picture from the Hubble Space Telescope shows the impact sige of fragment G of Comet Shoemaker-Levy 9 (SL9) on Jupiter's cloud tops. The dark outer ring of the impact site is approximately the size of the planet Earth. Photo credit for this images goest to H. Hammel (MIT), WFPC2, HST, and NASA.

sn1987a_heritage_big.jpg
The explosion of supernova sn1987a in the Large Magellanic Cloud has left rings and glowing gas that are being excited by light from the initial explosion. SN1987a is some 150,000 light-years distant; the remnants appear near the center of this Hubble Space Telescope image. Photo Credit: Hubble Heritage Team (AURA/ STScI/ NASA).

sn94d_hiz_big.jpg

The galaxy NGC4526 (apparent magnitude 9.6) with Supernova 1994D (the bright star at lower left in this image), as seen by the Hubble Space Telescope in this May 9, 1994 exposure. Photo credit: High-Z Supernova Search Team, HST, and NASA.

southerndeep_hst_big.jpg

A Hubble Space Telescope "deep" image of a small portion of the southern hemisphere sky, in this case in the constellation Tucana. This image is composited from many exposures taken over a ten-day period. Credit: R. Williams (STScI), HDF-S Team, and NASA.

sts076_715_mir.jpg

Like some giant mechanical insect, the Russian Space Station Mir floats above the planet Earth. The crew of the Space Shuttle during the STS 76 mission took this photograph in March 1996. Photo credit: the STS 76 Crew and NASA.

sts076_719_053.jpg

Cairo, Egypt and the Nile Delta at Lat/Lon: 30.5N/31.5E, was photographed by the Space Shuttle crew during the STS 76 mission in March 1996. Photo credit: the STS 76 Crew and NASA.

sts81atl.jpg

A high resolution view of the docked Space Shuttle Atlantis, taken from the Russian Space Station Mir. Photograph credit: NASA.

sunrise_sts47.gif

A sunrise from space, photographed by the Space Shuttle during the STS 47 mission. The flattened tops of the anvil-shaped storm clouds visible on the horizon mark the top of Earth's troposphere. Suspended dust causes this dense atmospheric layer to appear red, while the stratosphere appears as a blue stripe. Photo credit for this image goes to the STS 47 Crew and NASA.

sunspot_nso_big.jpg
Sunspots such as these are cooler magnetic depressions on the Sun's surface that typically last a month or two before dissipating. The number of sunspots varies, going from a minimum to a maximum about every 5.5 years. This photograph is credited to the National Solar Observatory.

sunspot_vtt.jpg
This close-up image of a sunspot also shows the bubbling cells of hot gas that covers the Sun's surface. These bubbles, called granules, are about 1,000 km across and typically last about ten minutes before exploding.

sunview_sts82_big.jpg
Astronaught Joseph Tanner, as photographed by Gregory Harbaugh during the second Hubble Space Telescope repair mission (STS-82) in February 1997. In the lower right corner of the image is the aft section of the Space Shuttle Discovery, and in the background is the sun and the crescent earth. Credit: STS-82 Crew and NASA.

titan_vg2_big.jpg
Saturn's moon Titan is one of only two moons in the Solar Systsem with an atmosphere—the other one is Neptune's Triton. The atmosphere of Titan consists primarily of nitrogen, but also contains chemicals such as methane and ethane. The Voyager 2 spacecraft produced this image in 1980 during its flyby of Saturn. The Cassini probe, launched in 1997, is expected to map Titan's surface in 2004. Credit: Voyager 2 and NASA.

titania_vg2.gif
Titania, Uranus' largest moon, as photographed by the Voyager 2 spacecraft during its 1986 flyby. Tita. nia is composed of about half water ice and half rock. At some point in history, Titania has obviously undergone some major resurfacing event, obvious from its jumbled terrain. This photograph is credited to NASA, Voyager 2, and Calvin J. Hamilton.

triton1_vg2_big.gif

Triton, Neptune's largest moon, as photographed by Voyager 2 in 1989. On Triton, Voyager 2 found a thin atmosphere and evidence of ice volcanoes. Photo credit: Voyager 2 and NASA.

uranus_vg2.gif

The only spacecraft to ever visit Uranus was Voyager 2, which transmitted back this image taken during its 1986 flyby. Uranus' rotation axis is greatly tilted, even pointing near the Sun at times—the reason for this axial tilt remains a mystery. Uranus presents a fairly bland surface, unlike the planets on either side of it—Saturn and Neptune. Image credit: NASA and Voyager 2.

usanight_dmsp_big.gif

The U.S.A. at night, seen by artificial light. This composite photograph was created from over 200 images made by Earth-orbiting satellites. Viewing this image, it is easy to understand why light pollution is a major concern for both amateur and professional Astronomers. Photo credit: the NOAA/NGDC/DMSP Digital Archive.

venera.gif

The surface of Venus appears from this photograph transmitted back to Earth from the Russian Venera spacecraft. Venera was only able to send a limited amount of data before failing under the incredibly hot and harsh conditions found on Venus.

vom_nj05s070.gif

Looking much like Earth's Grand Canyon, this image is an oblique color mosaic of Candor Chasma on the planet Mars, comprised of images from Viking Orbiter's 1 and 2. This view of the 800 kilometer wide Candor Chasma, part of the Valles Marineris system, is looking north. Both the walls and floor of this huge chasm show evidence of erosion, mass wasting, and complex geomorphology. Photo credit: Viking 1 and 2, NASA.

watervapor_goes8_big.gif

Although this might appear to be some exotic world from a science fiction movie, it is actually a view of planet Earth in a false color infrared image made by the Goestationary Operational Environmental Satellite (GOES). The instruments aboard GOES can produce images at the infrared wavelength of 6.7 microns, thus recording radiation emitted by water vapor in Earth's upper atmosphere. Bright areas correspond to high concentrations of water vapor, while dark areas are relatively dry. This unusual image is credited to F. Hasler et al., (NASA/GSFC) and The GOES Project.

SPECIAL: saganstationii.mov

A quicktime panorama of the Mars Pathfinder Landing site, which has been officially renamed as The Dr. Carl Sagan Memorial Station.

APPENDIX C

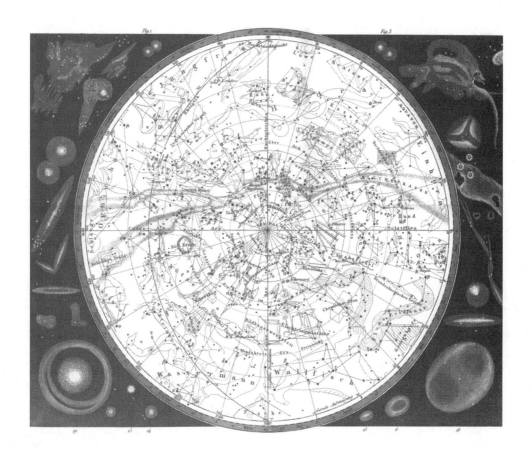

ANIMATIONS ON THE CD-ROM

Listed below are the animations of astronomical objects and phenomena included on the CD-ROM, in the **animations** sub-directory. (Sources are listed in parentheses.) The animations are available in a number of formats, including AVI, MPEG, and QuickTime. Many Web browsers already come bundled with the plug-in needed to view the files, but if your computer doesn't have the software necessary to view AVI or MPEG files (QuickTime is included on the CD-ROM), check the list of viewer software available at http://graffiti.u-bordeaux.fr/MAPBX/roussel/fra/softe.html. For information on how to view QuickTime VR panoramas (for the Mars Pathfinder files below) check http://quicktime.apple.com.

Asteroids

Gaspra2.qt
Movie showing the flyby of the asteroid Gaspra by NASA's Galileo spacecraft in 1991 (NASA/JPL).

Tout2.mpg
Animation of the rotation of the contact binary asteroid Toutatis (JPL).

Eclipses

ECL051094C.mpg
A movie of images from a weather satellite showing the progression of the shadow of the annular eclipse on May 10, 1994 (NOAA).

ECL071191.mpg
A movie of images from a weather satellite showing the progression of the shadow of the total eclipse on July 11, 1991 (NOAA).

ECL110394.mpg

Animation of the total solar eclipse of November 3, 1994, as seen from Bolivia (Fred Espenak, NASA Goddard Space Flight Center).

Jupiter

Io.avi

A movie showing the rotation of Io, one of the Galilean satellites of Jupiter (NASA Goddard Space Flight Center).

Io1.avi

A movie showing the rotation of Io, focusing on the volcanoes on its surface (NASA/JPL).

Jupiter.qt

A movie showing the rotation of Jupiter as seen by the Voyager spacecraft (NASA/JPL).

Jupiter_160.Mov

An animation showing the view of Jupiter from Io over the course of one day (NASA).

Jupiter1.avi

A movie of the rotation of the entire planet, focusing on the movement of different regions of the planet's atmosphere using Voyager images (NASA/JPL).

Jupiter3.avi

An animation of the planet's Great Red Spot, using Voyager images (NASA/JPL).

Mars

Mars_Hubble.mov

A movie showing the rotation of Mars, based on images from the Hubble Space Telescope (Space Telescope Science Institute).

Valles.mpg

A computer-generated animation of a flight through the giant Valles Marineris canyon on Mars (Marc Stengel).

Moon

Earth_Moon_Galileo.qt

The Earth-Moon system seen by Galileo (NASA/JPL).

Neptune

Nept96.mpg

A movie of the rotation of Neptune, using images taken in multiple wavelengths from the Hubble Space Telescope (Lawrence Sromovsky, University of Wisconsin, and the Space Telescope Science Institute).

Neptune.avi

The movie showing the rotation of the planet as seen by Voyager 2 in 1989 (NASA/JPL).

Triton2.avi

A computer-generated flyover of the surface of Triton, Neptune's largest moon, based on Voyager 2 images (NASA/JPL).

Pathfinder

Big_Stereo.mov

A QuickTime VR panorama of the Pathfinder landing site, which has been officially renamed the Dr. Carl Sagan Memorial Station (NASA/JPL). To view the image in 3-D you'll need a pair of red/blue 3-D glasses.

Minster.mov

A small color panorama of the landing site, in QuickTime VR (NASA/JPL).

Saganstationii.mov

A high-resolution color panorama (the "Presidential Panorama") of the Pathfinder landing site, in QuickTime VR (NASA/JPL).

Pluto

Pluto.mpg

A movie showing the rotation of Pluto using computer-enhanced images from the Hubble Space Telescope (Alan Stern, Southwest Research Institute, and the Space Telescope Science Institute).

Saturn

Csprbld2.mpg

An animation of the landing of the Huygens probe on the surface of Titan, Saturn's largest moon (NASA/JPL).

Cstitan2.mpg

An animation of a journey over the surface of Titan, ending at the landing site of the Huygens probe (NASA/JPL).

Cstrjct2.mpg

A movie showing the trajectory the Cassini spacecraft is following to get from Earth to Saturn (NASA/JPL).

Encelad.avi

A computer-generated flight over the surface of Enceladus, a moon of Saturn, using images from the Voyager spacecraft (NASA/JPL).

Saturn_Fly-By.qt

A movie showing Voyager 2's flyby of Saturn in 1981 (NASA/JPL).

Saturnstorm.mpg

A movie of a large storm seen in Saturn's atmosphere in 1990 by the Hubble Space Telescope (NASA).

Titan.mpg

A movie of infrared images of the surface of Titan, taken by the Hubble Space Telescope (Peter Smith, University of Arizona, and the Space Telescope Science Institute).

Shoemaker-Levy 9

Approach.mpg

A computer-generated animation of the fragments of Comet Shoemaker-Levy 9 as it approached Jupiter (ESA/ISD Visulab).

Fragr-Keck.mpg

The collision of Shoemaker-Levy 9 fragment R with Jupiter on July 21, 1994 (W. M. Keck Observatory/Imke de Pater, James R. Graham, Garrett Jernigan UC-Berkeley).

Impactsim-Lowell.mpg

A computer-generated animation of the collision of fragments of the comet with Jupiter (John Spencer, Lowell Observatory).

Sun

Hudson_Loops.mpg

A movie of images from the Yohkoh x-ray satellite showing a flare erupting on the surface of the Sun (NASA).

Uranus

Miranda.avi

A computer-generated flight over the rugged terrain of Miranda, using images collected during Voyager 2's flyby in 1986 (NASA/JPL).

Venus

Flight1.avi

A computer-generated flight over the surface of the Alta Regio region of Venus, using data collected by the Magellan spacecraft (NASA/JPL).

Flight3.avi

A narrated computer-generated flight over the surface of the Alpha Regio region of Venus, using data collected by the Magellan space-craft (NASA/JPL).

Venus.mpg

A rotating globe of Venus showing the topography of the planet (low areas in blue, high areas in red), using data collected by the Magellan spacecraft (NASA).

Venusfly.mpg

A view of the globe of Venus using images from Magellan, followed by a computer-generated flyover of one region of the planet (NASA/JPL).

APPENDIX D

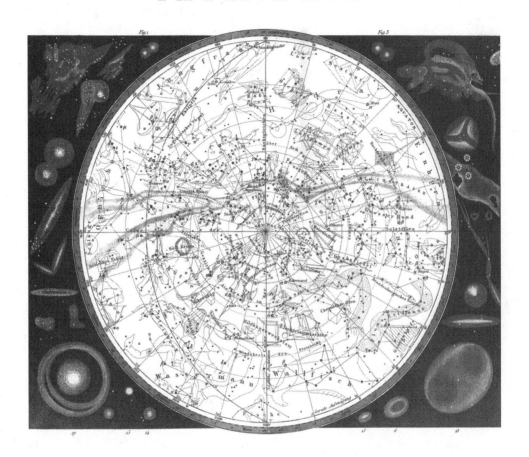

SOURCES

Telescopes & Accessories

Astronomics
2401 Tee Circle, Suites 105/106
Norman, OK 73069
Sales: (800) 422-7876
Phone: (405) 364-0858
Fax: (405) 447-3337
questions@astronomics.com
http://www.astronomics.com

Astro-Physics
11250 Forest Hills Road
Rockford, IL 61115
Phone: (815) 282-1513
Fax: (815) 282-9847

Celestron International
2835 Columbia Street
Torrance, CA 90503
Phone: (310) 328-9560
Fax: (310) 212-5835
http://www.celestron.com

Coulter Optical
(Part of MURNAGHAN
Instruments Corp.)
1781 Primrose Lane
West Palm Beach, FL 33414
Phone: (561) 795-2201
Fax: (561) 795-9889
murni@bix.com
http://www.murni.com

Jim's Mobile Inc.
810 Quail Street, Unit E
Lakewood, CO 80215
Sales: (800) 247-0304
Phone: (303) 233-5353
Fax: (303) 233-5359

Lumicon
2111 Research Drive #5
Livermore, CA 94550
Sales: (800) 767-9576
Phone: (925) 447-9570
Fax: (925) 447-9589

Meade Instruments Corporation
6001 Oak Canyon
Irvine, CA 92618-5200
Phone: (949) 451-1450
Fax: (949) 451-1460
http://www.meade.com

Orion Telescopes & Binoculars
P.O. Box 1815
Santa Cruz, CA 95061
Sales: (800) 447-1001
Phone: (831) 763-7000
Fax: (831) 763-7017
http://www.oriontel.com

Parks Optical
P.O. Box 716
Simi Valley, CA 93062
Phone: (805) 522-6722
Fax: (805) 522-0033

Questar Corporation
6204 Ingham Road
New Hope, PA 18938
Sales: (800) 247-9607
Phone: (215) 862-5277
Fax: (215) 862-0512
questar@erols.com
http://www.questar-corp.com

Software Bisque
912 Twelfth Street
Golden, CO 80401
Sales: (800) 843-7599
Phone: (303) 278-4478
Fax: (303) 278-0045
http://www.bisque.com

TeleVue
100 Route 59
Suffern, NY 10901
Phone: (914) 357-9522
http://www.televue.com

Thousand Oaks Optical
P.O. Box 4813
Thousand Oaks, CA 91359
Sales: (800) 996-9111
Phone: (805) 491-3642
Fax: (805) 491-2393

Roger W. Tuthill, Inc.
11 Tanglewood Lane
Mountainside, NJ 07092
Sales: (800) 223-1063
Phone: (908) 232-1786
Fax: (908) 232-3804

Unitron, Inc.
P.O. Box 469
Bohemia, NY 11716
Phone: (516) 589-6666
Fax: (516) 589-6975
info@unitronusa.com
http://www.unitronusa.com

University Optics
P.O. Box 1205
Ann Arbor, MI 48106
Sales: (800) 521-2828
Phone: (734) 665-3575
Fax: (734) 665-1815
http://www.
     universityoptics.com

Books

Astronomical Society of the Pacific
390 Ashton Avenue
San Francisco, CA 94112
Sales: (800) 335-2624
Phone: (415) 337-1100
Fax: (415) 337-5205
catalog@apsky.org
http://www.aspsky.org

Astronomy Book Club
Dept. L-EA7
300 Cindel Drive
Delran, NJ 08370-0001

Sky Publishing
49 Bay State Road
Cambridge, MA 02138
Sales: (800) 253-0245
Phone: (617) 864-7360
Fax: (617) 864-6117
http://www.skypub.com

Willmann-Bell, Inc.
P.O. Box 35025
Richmond, VA 23235
Sales: (800) 825-7827
Phone: (804) 320-7016
Fax: (804) 272-5920
http://www.willbell.com

Magazines

Astronomy Magazine
Kalmbach Publishing Co.
P.O. Box 1612
Waukesha, WI 53187-1612
Sales: (800) 533-6644
Phone: (414) 796-8776
Fax: (414) 796-1615
astro@astronomy.com
http://www.astronomy.com

Sky & Telescope Magazine
Sky Publishing
49 Bay State Road
Cambridge, MA 02138
Sales: (800) 253-0245
Phone: (617) 864-7360
Fax: (617) 864-6117
http://www.skypub.com

CCDs

Santa Barbara Instrument Group
P.O. Box 50437
Santa Barbara, CA 93150
Phone: (805) 969-1851
Fax: (805) 969-4069
sbig@sbig.com
http://www.sbig.com

Software

Luis Argüelles
http://members.tripod.
  com/~whuyss/mls.htm
whuyss@tripod.net
Messier Logging System software

David Chandler Company
P.O. Box 999
Springville, CA 93265
Phone: (559) 539-0900
Fax: (559) 539-7033
David@DavidChandler.com
http://www.davidchandler.com
Deep Space, The Observer's Guide
to the Night Sky

Fabio Cavicchio
Fax: +39 0544 463589
msb@ntt.it
http://www.
  sira.it/msb/Astro_en.htm
  AstroArt Image Processing software

Farrier Software
368 South McCaslin Boulevard
PMB 153
Louisville, CO 80027
Phone: (303) 926-0975
sales@farriersoft.com
http://www.farriersoft.com
(Hubble Images CD-ROM)

GeoClock
2218 North Tuckahoe Street
Arlington, VA 22205-1964
Phone: (603) 241-2661
Fax: (603) 241-5809
BBS: (603) 241-6980
Joe@CompuServe.com
http://www.clark.net/
    pub/bblake/geoclock

Insanely Great Software
Attn: Adam Stein
126 Calvert Avenue E.
Edison, NJ 08820
Sales: (800) 319-5106
Phone: (616) 266-1630
Fax: (632) 632-1666)
igs@igsnet.com
Web: http://www.igsnet.com
WinWeather software

Locutus Codeware
P.O. Box 53586
984 West Broadway
Vancouver
BC, V5Z 1K0 Canada
info@locutuscodeware.com
http://www.locutuscodeware.com
WetSock, SocketWatch,
MoonPhase, etc.

LogSat Software Corporation
425 Sout Chickasaw Tr., Suite 103
Orlando, FL 32825
Sales: (800) 350-3871
Phone: (407) 275-0780
Fax: (407) 275-2416
sales@logsat.com
http://www.logsat.com

Maraj Inc.
861 Pine Ridge Drive
Stone Mountain, GA 30087
Phone: (770) 413-5838
Fax: (770) 413-5838
maraj@bigfoot.com
http://www2.gsu.edu/~usgnrmx/
    maraj/products/prod01.html
    Planet Guide software, etc.

Maris Multimedia, Ltd.
99 Mansell Street
London, England E1 8AX
Phone: +44 (0) 171-488-1566
Fax: +44 (0) 171-702-0534
 RedShift 3

Network Cybernetics Corporation
4201 Wingren Road, Suite 202
Irving, TX 75062-2763
Phone: (972) 650-2002
Fax: (972) 650-1929
info@ncc.com
http://www.ncc.com/cdroms/
    index.html
Astroware 1 and SL9: Impact

Nova Astronomics
P.O. Box 31013
Halifax, Nova Scotia
B3K 5T9 Canada
Phone: (902) 499-6196
Fax: (902) 826-7957
info@nova-astro.com
http://www.nova-astro.com
Earth Centered Universe, Windows
Ephemeris Tool, etc.

Gary Nugent
54a Landscape Park
Churchtown
Dublin 14, Ireland
gnugent@indigo.ie
http://indigo.ie/~gnugent/JupSat95
JupSat95

Personal MicroCosms/Pocket Sized
Software
8547 East Arapahoe Road
PMB J-147
Greenwood Village, CO 80112
EricTerrell@juno.com
http://www.ericterrell.simplenet.com/
    html/ss_ac32.html
Astronomy Lab, etc.

Pickering Anomalies
support@anomalies.com
http://www.anomalies.com/
    asteroid/info.htm
Asteroid Pro software, etc.

Project Pluto
168 Ridge Road
Bowdoinham, ME 04008
Sales: (800) 666-5886
Tel.: (207) 666-5750
Fax: (207) 666-3149
pluto@projectpluto.com
http://www.projectpluto.com
Guide Software

Rainman Software
P.O.Box 6074
Gloucester, MA 01930-6174
Sales: (888) 261-1686
Fax: (888) 995-9191
ngcview@rainman-soft.com
http://www.rainman-soft.com
NGCView software

David H. Ransom, Jr.
240 Bristlecone Pines Road
Sedona, Arizona 86336
BBS: (520) 282-5559
rans7500@spacelink.nasa.gov
http://www.dransom.com

Retsik Software
P.O. Box 50
Enola, PA 17025-0050
Fax: (717) 732-5048
BBS: (717) 732-4067
70254.2017@compuserve.
    com or retsik@aol.com
http://www.retsik-software.com
 (AccuSet software)

RomTech, Inc.
2000 Cabot Boulevard West,
Suite 110
Langhorne, PA 19047-1811
Phone: (215) 750-6606
Fax: (215) 750-0803
http://www.romt.com
Universe Explorer: Distant Suns,
Mars Rover, etc.

SkyMap
9 Severn Road
Culcheth, Cheshire WA3 5ED UK
Phone: +44 (0) 1925 764131
sales@skymap.com
http://www.skymap.com
U.S. orders:
World Wide Software Publishing
P.O. Box 202036
Bloomington, MN 55420
Phone: (612) 881-7045
FAX: (612) 881-7024
Sales: (877) 889-0588
order@wwsoftware.com
http://www.wwsoftware.com

Software Bisque
912 Twelfth Street, Suite A
Golden, CO 80401
Sales: (800) 843-7599
Phone: (303) 278-4478
Fax: (303) 278-0045
http://www.bisque.com

Southern Stars Software
P.O. Box 3792
Saratoga, CA 95070
Phone: (408) 293-2362

info@southernstars.com
http://www.southernstars.com

Steven S. Tuma
1425 Greenwich Lane
Janesville, WI 53545
Phone: (608) 752-8366
stuma@inwave.com
http://www.deepsky2000.com
DeepSky 99 software

Paul E. Traufler
111 Emerald Drive
Harvest, AL 35749
Phone: (256) 837-0084
Fax: (256) 895-0654
wintrak@traveller.com
http://www.hsv.tis.net/~wintrak

Willmann-Bell, Inc.
P.O. Box 35025
Richmond, VA 23235
Sales: (800) 825-7827 (orders only)
Phone: (804) 320-7016 (orders and
information)
Fax: (804) 272-5920
http://www.willbell.com
MegaStar Deep Sky Atlas, MICA

Miscellaneous

Edmund Scientific
101 E. Gloucester Pike
Barrington, NJ 08007-1380
Phone: (800) 728-6999
Fax: (856) 547-2392
info@edsci.com
http://www.edsci.com

Astronomy Clubs, Planetariums, Science Museums, and Observatories:

ALPO (International)
http://www.lpl.arizona.edu/~rhill/alpo/clublinks.html

Astonomy-Mall.com (International)
http://astronomy-mall.com/ regular/clubs-etc/clubsetc.html

Astronomical Society of South Australia (Australia)
http://www.assa.org.au

Astronomy Magazine's Club Listings *
http://www2.astronomy.com/astro
(or)
http://www.kalmbach.com/astro/Hobby/Hobby.html

Astronomical Society of the Pacific (International)
http://www.aspsky.org/html/resources/ amateur.html

Astronomy: Amateur
http://webhead.com/WWWVL/Astronomy/amateur.html

Astroplace (International)
http://astroplace.com/amateurs/clubs.htm

Expanding Universe (International)
http://www.mtrl.toronto.on.ca/centres/bsd/astronomy/520_6.HTM

Federation of Astronomical Societies (U.K.)
http://www.fedastro.demon.co.uk/societies/index.html

Répertoire des clubs en France (France)
http://perso.club-internet.fr/uranos/clubfr.htm

Sky & Telescope Magazine's Directory Listings *
http://www.skypub.com/astrodir/astrodir.html
(or)
http://www.skypub.com/links/clubsweb.html

StarWorlds (searchable directory, International):
http://cdsweb.u-strasbg.fr/~heck/sfworlds.htm

U.K. Amateur Astronomy (U.K.)
http://www.ukindex.co.uk/ukastro/refssocs.html

Under the Western Sky (western U.S.)
http://www.rtd1.com/westernsky

Yahoo science, astronomy (International)
http://www.yahoo.com/Science/Astronomy/Clubs

* Recommended Resource

UseNet News Groups

alt.binaries.pictures.astro
alt.sci.planetary
sci.astro
sci.astro.amateur
sci.astro.ccd-imaging
sci.astro.planetarium

Index

Piazzi, Giuseppi, 62
Pictor (CCD camera), 283
Pike, John, 48
PixCel (CCD camera), 283
Pixel Vision Web site, 284
PKS 2349 (quasar), *190*, 191
Plait, Phil, 25
Planck (spacecraft), 208
PLANET 1A (software), 359
planetarium programs. *See also*
    educational software; observa-
    tional software
    about, 213
    on CD-ROM with this book,
        311–312, 332–340, 370, 374
    comet information in, 74–75
    examples of, 213–225
    for Macintosh, 370, 374
    Moon information in, *32–33*
    simulated views of Earth in,
        21–22
    star information in, 111
    Sun information in, 86, 88
    uses for, 3–4
    using, to control telescopes,
        278–279
    using, to find lunar eclipses, 35,
        36
    using, to identify objects in the
        sky, 3–5, 225
    using, to locate galaxies,
        164–165
    using, to observe asteroids,
        64–65
planetariums, 253, 443–444
Planetary Missions (software), 236,
    237

planetary nebulae. *See also* deep
    sky object catalog codes for
    individual objects
    about, 124, 172–173
    Cat's Eye, 397
    databases of, 173
    images of, 173
    images on the CD-ROM with
        this book, 387–389, 391,
        397, 398–399, 409, 414, 415
Planetary Nebulae Sampler Web
    site, 173
Planetary Photojournal, 41
planetary science newsgroups,
    253–254
Planetary Society Web page, 86, *87*
Planetary System, The (software),
    42
Planet Guide (software), 242, *243*,
    360
PlanetQuest (software), 42
planets. *See also* names of individual
    planets; solar system
    around other suns, 113–*117*
    information about, on AOL,
        247–248
    observing, 287
    software programs about, 236,
        237, 350–363, 377–380
Planets, The (software), 236, *237*
PLANETS (software), 360
PlanetWatch (software), 361
plasmas, formation of, 205
Pleiades
    images of, 409
    nebula around, 171, *172*
    Web site, 176

Pluto
    about, 59
    animations, 432
    images of, 58, *59*, 386
    information sources, 57–59
    locating in the sky, 43
    spacecraft exploration of, 58–59
Pluto-Kuiper Express (spacecraft),
    58–59
Polaris (star), *112*
press releases, 254, 255
Princeton University
    cosmologists' research at, 201
    Web site, 159, *160*
Prodigy (online service), 250, *250*
Project CLEA software, 242, *243*,
    303, 318, 350
Proxima Centauri, 394–395
PSR1257+12 (pulsar), 114
pulsar planets, discovery of,
    114–117
pulsars
    about, 138–140
    images of, 140–*141*, 399
    information sources, 140

Q

QSOs (quasi-stellar objects), 189
quarks, 197
quasars
    and black holes, 191–192
    catalogs of, 186
    definition of, 189
    discovery of, 188–189
    and galaxies, 186, 189, 192
    galaxies having, 162
    information sources, 189

PLAYING LINUX GAMES

by AL KOSKELIN

The first complete guide to gaming for Linux users, *Playing Linux Games* covers Linux basics as well as game-specific hardware and software issues. Linux can be a great gaming platform and this book has everything you'll need to get games up and running fast, including dozens of games, drivers, and information files on the bundled CD-ROM. You'll learn:

- How to find, install, run, and troubleshoot games on Linux

- How to get the software or drivers you need to play games under Linux

- How to set up a LAN to play games over a network

- How to install and play the games currently available for Linux (including games for other operating systems that can be emulated on Linux)

- Game-specific tips, techniques, and strategy

AL KOSKELIN studies computer science and writing at the University of Wisconsin-Stevens Point. He is the co-founder of the popular Linux Games Web site (http://www.linuxgames.com).

350 pp., paperback, $34.95 w/CD-ROM
ISBN 1-886411-33-6

LINUX MIDI & SOUND

by DAVE PHILLIPS

Linux MIDI & Sound offers in-depth instruction on recording, storing, playing, and editing music and sound under Linux. The author, a programmer and performing musician, discusses the basics of sound and digital audio, and covers specific software and hardware issues specific to Linux, including:

- A clear introduction to the fundamental concepts of digital sound

- Linux-specific issues including available toolkits, GUI libraries, and driver support

- Reviews of available software with recommendations

- Recommended components for building a complete system including a digital audio player/recorder, soundfile editor, MIDI recorder/player/editor, and software mixer

- Coverage of hard disk recording, advanced MIDI support, network audio, and MP3

- A complete bibliography and an extensive list of Internet resources

- A CD-ROM with dozens of software packages

A performing musician for over 30 years, DAVE PHILLIPS became interested in computers as a means for playing, editing, and recording music. He is an expert in MIDI, Csound, and Linux. He currently maintains several educational Web sites on these topics.

300 pp., paperback, $39.95 w/CD-ROM
ISBN 1-886411-34-4

APPLE CONFIDENTIAL:
THE REAL STORY OF APPLE COMPUTER, INC.

by OWEN W. LINZMAYER

"The Apple story . . . in all its drama." — *The New York Times Book Review*

"Written with humor, respect, and care, it absolutely is a must-read for every Apple fan." — *InfoWorld*

"An irreverent work that captures the essence of that which is Apple."
— *Gil Amelio, former Apple CEO*

In *Apple Confidential*, journalist and Mac guru Owen Linzmayer traces the history of Apple Computer from its legendary founding to its recent return to profitability. Backed by exhaustive research, the book debunks many of the myths and half-truths surrounding Apple, the Macintosh, and its creators. With facts, photos, timelines, and quotes, every page brings a new discovery about this fascinating company, including:

- The forgotten founder who walked away from hundreds of millions of dollars

- The trials and tribulations of taking the original Macintosh from research project to finished product

- Apple's disastrous oversight that allowed Microsoft to dominate the personal computing industry

- The careers of CEOs Sculley, Spindler, Amelio, and Jobs

- Complete timelines of the development of the Apple II, Apple III, Lisa, Macintosh, Newton, NeXT, and Windows

- The triumphant return of Steve Jobs to Apple

OWEN W. LINZMAYER is a San Francisco-based freelance writer who has been covering Apple for industry magazines since 1980. He writes a monthly column for *MacAddict* magazine and is the author of four other Macintosh books, including *The Mac Bathroom Reader*.

268 pp., paperback, $17.95
ISBN 1-886411-28-X

STEAL THIS COMPUTER BOOK: WHAT THEY WON'T TELL YOU ABOUT THE INTERNET

by WALLACE WANG

"A delightfully irresponsible primer." — *Chicago Tribune*

"If this book had a soundtrack, it'd be Lou Reed's 'Walk on the Wild Side.'" — *InfoWorld*

"An unabashed look at the dark side of the Net — the stuff many other books gloss over." — *Amazon.com*

Steal This Computer Book explores the dark corners of the Internet and reveals little-known techniques that hackers use to subvert authority. Unfortunately, some of these techniques, when used by malicious hackers, can destroy data and compromise the security of corporate and government networks. To keep your computer safe from viruses, and yourself from electronic con games and security crackers, Wallace Wang explains the secrets hackers and scammers use to prey on their victims. Discover:

- How hackers write and spread computer viruses

- How criminals get free service and harass legitimate customers on online services like America Online

- How online con artists trick people out of thousands of dollars

- Where hackers find the tools to crack into computers or steal software

- How to find and use government-quality encryption to protect your data

- How hackers steal passwords from other computers

WALLACE WANG is the author of several computer books, including *Microsoft Office 97 for Windows for Dummies* and *Visual Basic for Dummies*. A regular contributor to *Boardwatch* magazine (the "Internet Underground" columnist), he's also a successful stand-up comedian. He lives in San Diego, California.

340 pp., paperback, $19.95
ISBN 1-886411-21-2

If you can't find No Starch Press titles in your local bookstore, here's how to order directly from us (we accept MasterCard, Visa, and checks or money orders — sorry, no CODs):

PHONE:

1 (800) 420-7240 OR
(415) 863-9900
MONDAY THROUGH FRIDAY,
9 A.M. TO 5 P.M. (PST)

FAX:

(415) 863-9950
24 HOURS A DAY,
7 DAYS A WEEK

E-MAIL:

SALES@NOSTARCH.COM

WEB:

HTTP://WWW.NOSTARCH.COM

MAIL:

NO STARCH PRESS
555 DE HARO STREET, SUITE 250
SAN FRANCISCO, CA 94107
USA

Distributed to the book trade by Publishers Group West

Using the Electronic Book

Included on the CD-ROM bundled with this book (filename: **Book.pdf** in the "Book" subdirectory) is a *PDF* (Portable Document Format) file—an easily navigable electronic file—of the *Astronomer's Computer Companion*. You can quickly and easily navigate through the PDF file of the book by clicking chapter titles or subheadings in the PDF's table of contents or bookmarks.

Installing Adobe Acrobat Reader

To view the PDF file of the *Astronomer's Computer Companion*, you must have the Adobe Acrobat Reader installed. Versions of Adobe's Acrobat Reader for Macintosh and Windows are included on the CD-ROM in the Software directory (PC/ADOBE and Mac/Adobe). To install, click the appropriate file (**ar40eng.exe** for Windows 95 or higher, **1r16e301.exe** for Windows 3.1, and **ar40eng.sit.hqx** for Macintosh) and follow the on-screen instructions. See page 371 for Macintosh Operating System requirements.

Navigating through the Electronic Book

The easiest way to navigate through the PDF file of the book is to use the bookmarks. You can open the bookmarks by choosing **Window | Show Bookmarks** or by pressing **F5**.

Clicking on a chapter title or subheading in the bookmarks will take you to that page; you can then scroll through the text. You can also use the book's Table of Contents to quickly link to a particular section of the book.

To find a specific page, use the thumbnails window (**Window | Show Thumbnails**), to display thumbnail sketches of each book page. Double-click a thumbnail to view that page.

Note: The page numbers underneath the thumbnails do not correspond to the actual book's page numbers.

Acrobat Reader also lets you search for words or phrases in PDF documents. To search, choose **Edit | Find** and enter a search term or phrase.

Web, Image, Software, and Animation Links

All the web sites listed in the text of the *Astronomer's Computer Companion* are live links (coded blue) in the PDF file. To visit them, click their address to launch the web site in your browser.

The animations and images on the CD-ROM are coded red. Clicking a red link launches the corresponding image or animation in your image or movie viewer.

Note: You must have a viewer installed for these image links to function. VuePrint (included on the CD-ROM in Software/PC/Vueprint) will let you view all the images and most of the animations, and QuickTime (Software/PC/Quicktime; or Software/Mac/Quicktime) will run the remaining animations (usually .MOV files). See Appendix A (page 367 for VuePrint, page 306 for Macintosh QuickTime, and page 306 for PC QuickTime) for specific installation instructions and Operating System requirements.

Some programs are also coded red. Clicking on these links will install the corresponding program.

Updates

This book was carefully reviewed for technical accuracy, but it's inevitable that some things will change after the book goes to press. Visit the Web site for this book at **http://www.nostarch.com/ acc_updates.htm** for updates, errata, and information about downloading packages that are not included on the CD.

About the Authors

Jeff Foust received his Ph.D. from MIT in planetary astronomy. He serves on the board of directors of the Boston chapter of the National Space Society and edits the SpaceViews Web site (http:// www.spaceviews.com).

Ron LaFon is a freelance writer, editor, and a computer graphics and electronic publishing specialist from Sonoma, California. He has been an avid amateur astronomer since age seven.

License Agreement for *Astronomer's Computer Companion*

Read this Agreement before opening this package. By opening this package, you agree to be bound by the terms and conditions of this Agreement.

This CD-ROM (the "CD") contains programs and associated documentation and other materials and is distributed with the book entitled *Astronomer's Computer Companion* to purchasers of the book for their own personal use only. Such programs, documentation and other materials and their compilation (collectively, the "Collection") are licensed to you subject to terms and conditions of this Agreement by No Starch Press, having a place of business at 555 De Haro Street, Suite 250, San Francisco, CA 94107 ("Licensor"). In addition to being governed by the terms and conditions of this Agreement, your rights to use the programs and other materials included on the CD may also be governed by separate agreements distributed with those programs and materials on the CD (the "other Agreements"). In the event of any inconsistency between this Agreement and any of the Other Agreements, those Agreements shall govern insofar as those programs and materials are concerned. By using the Collection, in whole or in part, you agree to be bound by the terms and conditions of this Agreement. Licensor owns the copyright to the Collection, except insofar as it contains materials that are proprietary to third party suppliers. All rights in the Collection except those expressly granted to you in this Agreement are reserved to Licensor and such suppliers as their respective interests may appear.

1. **Limited License.** Licensor grants you a limited, nonexclusive, nontransferable license to use the Collection on a single dedicated computer (excluding network servers). This Agreement and your rights hereunder shall automatically terminate if you fail to comply with any provision of this Agreement or the Other Agreements. Upon such termination, you agree to destroy the CD and all copies of the CD, whether lawful or not, that are in your possession or under your control. Licensor and its suppliers retain all rights not expressly granted herein as their respective interests may appear.

2. **Additional Restrictions.** (A) You shall not (and shall not permit other persons or entities to) directly or indirectly, by electronic or other means, reproduce (except for archival purposes as permitted by law), publish, distribute, rent, lease, sell, sublicense, assign, or otherwise transfer the Collection or any part thereof or this Agreement. Any attempt to do so shall be void and of no effect. (B) You shall not (and shall not permit other persons or entities to) reverse-engineer, decompile, disassemble, merge, modify, create derivative works of, or translate the Collection or use the Collection or any part thereof for any commercial purpose. (C) You shall not (and shall not permit others persons or entities to) remove or obscure Licensor's or its suppliers' or Licensor's copyright, trademark, or other proprietary notices or legends from any portion of the Collection or any related materials. (D) You agree and certify that the Collection will not be exported outside the United States except as authorized and as permitted by the laws and regulations of the United States. If the Collection has been rightfully obtained outside of the United States, you agree that you will not reexport the Collection, except as permitted by the laws and regulations of the United States and the laws and regulations of the jurisdiction in which you obtained the Collection.

3. **Disclaimer of Warranty.** (A) The Collection and the CD are provided "as is" without warranty of any kind, either express or implied, including, without limitation, any warranty of merchantability and fitness for a particular purpose, the entire risk as to the results and performance of the CD and the software and other materials that is part of the Collection is assumed by you, and Licensor and its suppliers and distributors shall have no responsibility for defects in the CD or the accuracy or application of or errors or omissions in the Collection and do not warrant that the functions contained in the Collection will meet your requirements, or that the operation of the CD or the Collection will be uninterrupted or error-free, or that any defects in the CD or the Collection will be corrected. In no event shall Licensor or its suppliers or distributors be liable for any direct, indirect, special, incidental, or consequential damages arising out of the use of or inability to use the Collection or the CD, even if Licensor or its suppliers or distributors have been advised of the likelihood of such damages occurring. Licensor and its suppliers and distributors shall not be liable for any loss, damages, or costs arising out of, but not limited to, lost profits or revenue; loss of use of the Collection or the CD; loss of data or equipment; cost of recovering software, data, or materials in the Collection; the cost of substitute software, data, or materials in the Collection; claims by third parties; or other similar costs. (B) In no event shall Licensor or its suppliers' or distributors' total liability to you for all damages, losses, and causes of action (whether in contract, tort or otherwise) exceed the amount paid by you for the Collection. (C) Some states do not allow exclusion or limitation of implied warranties or limitation of liability for incidental or consequential damages, so the above limitations or exclusions may not apply to you.

4. U.S. Government Restricted Rights. The Collection is licensed subject to RESTRICTED RIGHTS. Use, duplication, or disclosure by the U.S. Government or any person or entity acting on its behalf is subject to restrictions as set forth in subdivision (c)(1)(ii) of the Rights in Technical Data and Computer Software Clause at DFARS (48 CFR 252.227-7013) for DoD contracts, in paragraphs (c)(1) and (2) of the Commercial Computer Software Restricted Rights clause in the FAR (48 CFR 52.227 - 19) for civilian agencies, or, in the case of NASA, in clause 18-52.227-86(d) of the NASA Supplement to the FAR, or in other comparable agency clauses. The contractor/manufacturer is No Starch Press, 555 De Haro Street, Suite 250, San Francisco, CA 94107.

5. General Provisions. Nothing in this Agreement constitutes a waiver of Licensor's, or its suppliers' or licensors' rights under U.S. copyright laws or any other federal, state, local, or foreign law. You are responsible for installation, management, and operation of the Collection. This Agreement shall be construed, interpreted, and governed under California law. Copyright ©2000 No Starch Press. All rights reserved. Reproduction in whole or in part without permission is prohibited.

"Since 1994, No Starch Press has published computer books that make a difference. We hope that this book has made a difference for you."

—William Pollock, Publisher
billp@nostarch.com

no starch press
555 De Haro Street
Suite 250
San Francisco, CA 94107

e-mail: info@nostarch.com
web: http://www.nostarch.com
tel: 1.800.420.7240
fax: 1.415.863.9950

GIVE US A PIECE OF YOUR MIND

Which book did this card come from?

How could this book be improved?

Did this book meet your expectations?
Why/Why not?

Any suggestions for other computer books?

❏ **PLEASE ADD ME TO YOUR MAILING LIST**   ❏ E-MAIL ONLY

NAME: _____   COMPANY _____

ADDRESS: _____

CITY: _____   STATE: _____ ZIP: _____ COUNTRY: _____

PHONE: _____   E-MAIL: _____